DESIGN TECHNIQUES FOR INTEGRATED CMOS CLASS-D AUDIO AMPLIFIERS

ADVANCED SERIES IN ELECTRICAL AND COMPUTER ENGINEERING

Editor: W. K. Chen *(University of Illinois, Chicago, USA)*

Published:

Vol. 7: Diagnostic Measurements in LSI/VLSI Integrated Circuits Production
by A. Jakubowski, W. Marciniak and H. Przewlocki

Vol. 8: An Introduction to Control Systems (Second Edition)
by K. Warwick

Vol. 9: Orthogonal Functions in Systems and Control
by K. B. Datta and B. M. Mohan

Vol. 10: Introduction to High Power Pulse Technology
by S. T. Pai and Q. Zhang

Vol. 11: Systems and Control: An Introduction to Linear, Sampled and Nonlinear Systems
by T. Dougherty

Vol. 12: Protocol Conformance Testing Using Unique Input/Output Sequences
by X. Sun, C. Feng, Y. Shen and F. Lombardi

Vol. 13: Semiconductor Manufacturing Technology
by C. S. Yoo

Vol. 14: Linear Parameter-Varying System Identification: New Developments and Trends
by Paulo Lopes dos Santos, Teresa Paula Azevedo Perdicoúlis, Carlo Novara, Jose A. Ramos and Daniel E. Rivera

Vol. 16: Design Techniques for Integrated CMOS Class-D Audio Amplifiers
by Adrian I. Colli-Menchi, Miguel A. Rojas-Gonzalez and Edgar Sanchez-Sinencio

Vol. 18: Broadband Matching: Theory and Implementations, 3rd Edition
by Wai-Kai Chen

Forthcoming:

Active Network Analysis: Feedback Amplifier Theory — 2nd Edition
by Wai-Kai Chen (University of Illinois, Chicago, USA)

*For the complete list of titles in this series, please visit
http://www.worldscientific.com/series/asece

Advanced Series in Electrical and Computer Engineering – Vol. 16

DESIGN TECHNIQUES FOR INTEGRATED CMOS CLASS-D AUDIO AMPLIFIERS

Adrian I Colli-Menchi
Miguel A Rojas-Gonzalez
Edgar Sánchez-Sinencio

Texas A&M University, USA

World Scientific

NEW JERSEY • LONDON • SINGAPORE • BEIJING • SHANGHAI • HONG KONG • TAIPEI • CHENNAI • TOKYO

Published by

World Scientific Publishing Co. Pte. Ltd.
5 Toh Tuck Link, Singapore 596224
USA office: 27 Warren Street, Suite 401-402, Hackensack, NJ 07601
UK office: 57 Shelton Street, Covent Garden, London WC2H 9HE

Library of Congress Cataloging-in-Publication Data
Names: Colli-Menchi, Adrian I., author. | Rojas-Gonzalez, Miguel A. (Miguel Angel), author. | Sâanchez-Sinencio, Edgar, author.
Title: Design techniques for integrated CMOS class-D audio amplifiers / Adrian I Colli-Menchi (Texas A&M University, USA), Miguel A Rojas-Gonzalez (Texas A&M University, USA) & Edgar Sanchez-Sinencio (Texas A&M University, USA).
Description: [Hackensack?] New Jersey : World Scientific, [2016] | Series: Advanced series in electrical and computer engineering ; Volume 16 | Includes bibliographical references and index.
Identifiers: LCCN 2016015377 | ISBN 9789814704243 (hc : alk. paper) | ISBN 9789814699426 (pbk : alk. paper)
Subjects: LCSH: Audio amplifiers--Design and construction. | Metal oxide semiconductors, Complementary. | Sound--Recording and reproducing--Equipment and supplies. | Application-specific integrated circuits.
Classification: LCC TK7871.58.A9 C65 2016 | DDC 621.3815/35--dc 3
LC record available at https://lccn.loc.gov/2016015377

British Library Cataloguing-in-Publication Data
A catalogue record for this book is available from the British Library.

Desk Editor: V. Vishnu Mohan

Typeset by Stallion Press
Email: enquiries@stallionpress.com

Printed in Singapore

To my wife and children.
Adrian I. Colli-Menchi

To Lucio, Isabel, Raziel, Lucy, and Mateo.
Miguel Angel Rojas-Gonzalez

To Gael, Alexandra, and Mateo.
Edgar Sanchez-Sinencio

Foreword

During the last 20 years, class-D amplifiers have become the leading technology for implementing power stages in audio applications, due to their high efficiency, intrinsic linearity, robust implementation, and compatibility with advanced CMOS technologies. Indeed, the class-D audio amplifier market is growing rapidly, driven by widespread applications, such as smartphones, music players, and in general personal devices, in which reduced dimensions and low power consumption are of paramount importance.

Although ideally the operating principle of class-D amplifiers is pretty simple, their design is not at all straightforward, especially when high-fidelity (HiFi) audio performance is required, as I figured out the hard way while working for a long time in the field of audio amplifiers, both in collaboration with leading industries and within academic research projects. Designing high-performance class-D amplifiers, indeed, requires specific know-how and involves a lot of knacks that are not easy to attain.

This book provides a comprehensive overview of the concepts and the techniques behind the design of audio class-D amplifiers in CMOS technology, starting from the fundamentals on audio amplification, through the basic class-D amplifier architectures, the circuit topologies, the building blocks, the design trade-offs, and the control strategies, ending with the power-supply noise enhancement techniques and the class-D amplifiers for unconventional applications.

The material presented in *Design Techniques for Integrated CMOS Class-D Audio Amplifiers* is organized hierarchically, thus making it easy to read and useful both as a reference book in industry and as text book in academia. Indeed, the reader is led systematically through the design process of class-D amplifiers, from the specifications to the silicon implementation.

The authors of the book, Dr. Adrian Colli-Menchi, Dr. Miguel Rojas-Gonzalez, and Dr. Edgar Sanchez-Sinencio, have been designing class-D amplifiers for many years at Texas A&M University, with the support of several leading industries in this field, such as Texas Instruments, Silicon Labs, Qualcomm, and Broadcom, accumulating a significant amount of knowledge, know-how, and expertise, which has been transferred in this book.

From this book, young engineers approaching the field of class-D amplifiers can acquire all of the basic information, whereas expert designers can attain several helpful tips and tricks for specific problems.

In conclusion, *Design Techniques for Integrated CMOS Class-D Audio Amplifiers* by Adrian Colli-Menchi, Miguel Rojas-Gonzalez, and Edgar Sanchez-Sinencio is an extremely useful tool for anyone working in the field of audio class-D amplifier.

Piero Malcovati
Associate Professor
Department of Electrical, Computer,
and Biomedical Engineering
University of Pavia, Italy

Preface

The main objective of this book is to introduce the techniques and principles for the design of class-D amplifiers implemented in CMOS technology. This book was developed from several sources; it reflects, in part, the research work on class-D amplifiers in the Analog and Mixed-Signal Center (AMSC) at Texas A&M University (TAMU) from 2007–2015. Part of this material has been used to supplement several graduate courses at TAMU, and invited seminars at industry. Furthermore, this book should be useful both as a reference book in industry, and as textbook in senior or graduate course on the subject. It is assumed that the reader is familiar with basic CMOS analog design circuits, but if the reader needs to review in detail the basics of analog design, some references are included where is pertinent.

Class-D amplifiers provide great advantages as far as heat dissipation, space reduction, and energy efficiency which are determining factors for its fast evolution in the emerging markets. Currently, switching amplifiers are present in several applications like consumer electronics, automotive, and avionics systems. The lower power dissipation of class-D audio systems allows less heat, reduces cost and circuit board space, and extended battery life in portable systems. All these characteristics, make the class-D amplifier very attractive for future audio applications.

Most portable electronics, home audio/video devices, and the more and more car audio systems all are heavily using class-D audio amplifiers. The market of class-D audio amplifier, in the last few years, is growing at a rate of over 50 percent. Thus, the authors see of tremendous importance the class-D design knowledge, both in industry and academia. This book has been strongly influenced by the research support of Texas Instruments, Silicon Labs, Qualcomm, and Broadcom.

The organization of this book is illustrated at the end of the preface. Chapter 1 is an introduction to fundamentals of audio amplification. It covers the principles of sound, and of audio amplifiers as well as audio amplifier classification. Chapter 2 discusses the principles of class-D audio amplifiers, describes advantages and weakness of class-D amplifiers and typical applications. Chapter 3 introduces the main class-D amplifier topologies and their tradeoffs. Chapter 4 describes class-D circuit design techniques including their basic building blocks. Chapter 5 is focused on power-supply noise enhancement techniques in class-D amplifiers, the characterization, and solutions using feedforward techniques. In previous chapters the topologies described use pulse width, sigma delta, or self-oscillating modulation, in Chapter 6 a different class-D control strategy that implements sliding-mode control is described. Finally, Chapter 7 describes class-D amplifiers that do not use typical electromagnetic speakers but piezoelectric speakers; their challenges and solution design techniques are discussed.

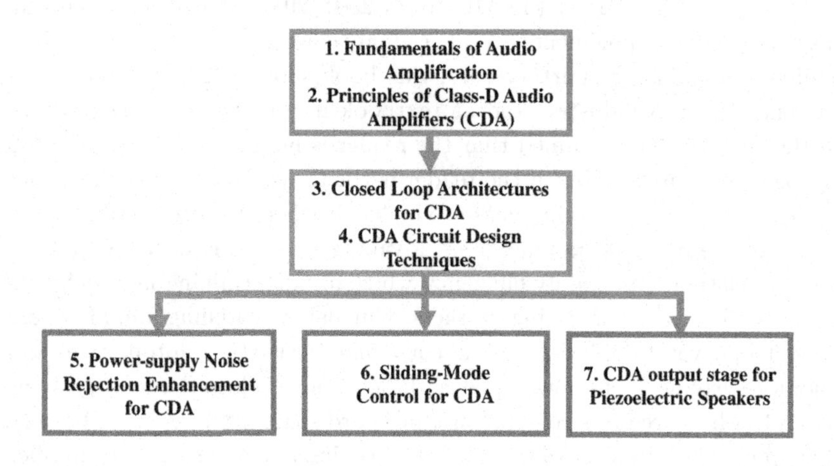

Acknowledgments

We want to thank all AMSC graduate students who collaborate directly or indirectly to the realization of this book, and in particular to Joselyn Torres for his great contribution and technical discussions. We are also grateful to the following colleagues for comments, suggestions, and corrections: Piero Malcovati, Gael Pillonnet, Douglas Lopata, Fernando Lavalle, Xiaosen Liu, Marco Berkhout, and Xicheng Jiang. We appreciate their time and effort spent while reviewing this book.

Adrian I. Colli-Menchi
Miguel Angel Rojas-Gonzalez
Edgar Sanchez-Sinencio

Contents

Chapter 1

Fundamentals of Audio Amplification

1.1 Introduction

The audio power amplifier provides the power to the loudspeaker in a sound system. In detail, the audio amplifier is a device that takes an input electrical signal representing the desired audio information, amplifies it, and delivers it to a transducer that converts the electrical signal back to audio as described in Fig. 1.1.

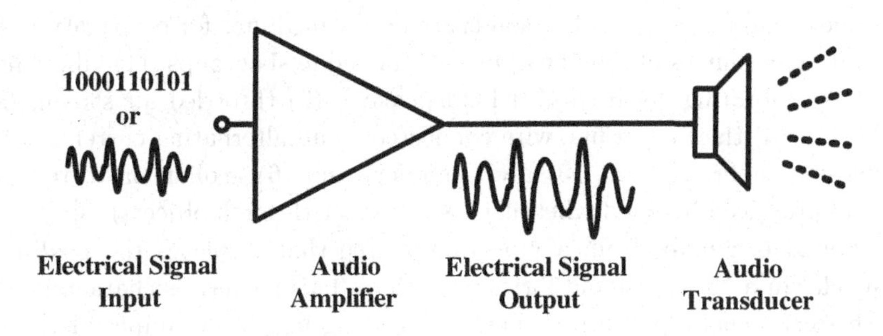

Fig. 1.1 Audio amplifier operation.

Nowadays most of portable devices in the market process the audio signal content using an audio processor. The audio processor sends the signal to the audio amplifier using a serial protocol such as Time-division Multiplexing (TDM), Pulse-density Modulation (PDM), or Inter-IC Sound (I2S). Most audio amplifiers include a serial interface to communicate with the audio processor. This communication capability allows added functions such as sound equalization, volume control, among others to the audio amplifier. The received audio signal is either converted to an analog signal using a Digital-to-Analog (DAC) converter or processed digitally to be applied to the audio amplifier. Note that the input of the audio amplifier can be continuous in time and amplitude (analog signal), or discrete in time and amplitude (digital signal), but the output signal applied to the audio transducer must be an analog signal.

Audio amplifiers have a history extending back to the 1900s, and with hundreds of thousands amplifiers being built nowadays, they are of considerable economic importance [1]. They can be found in many systems [2] such as televisions, radios, phones, cellular phones, hearing aids, portable audio/video players, tablets, desktop, notebook and netbook computers, car audio systems, home theater systems, and home audio systems. Therefore, there is a present and future need for high performance, efficient, and reliable audio power amplifiers in a host of applications.

1.2 Principles of sound and audio amplifiers

Unlike electromagnetic waves which propagate through free space, sound waves require a solid, liquid, or gas medium to propagate. Sound can be defined as a mechanical pressure wave moving in an elastic medium. Air is the most familiar medium but solids are better mediums for propagation of sound. The sources of sound can be categorized in six groups: (1) vibrating body (a vibrating vocal cord, a loudspeaker), (2) throttled air stream (a whistle), (3) thermal (a fine wire connected to an alternating current), (4) explosion (a firecracker), (5) arc (thunder) and (6) aeolian or vortex (a sound produced by wind when it passes over or through objects) [3].

Sound originates from a vibration source that displaces the medium particles in a backward and forward motion. This pattern is characterized with some generic properties such as wavelength, period, amplitude, and direction.

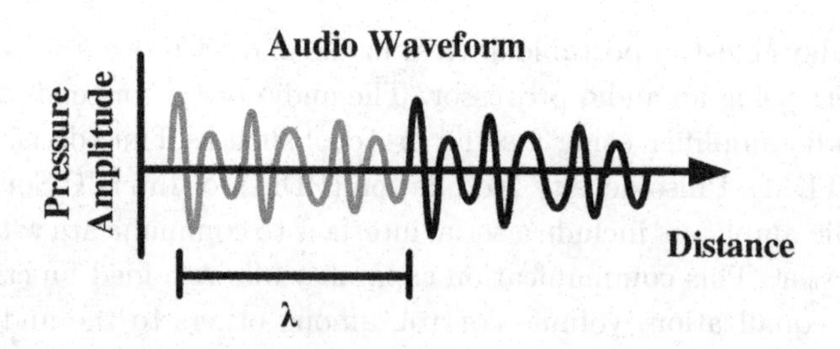

Fig. 1.2 Audio waveform across distance.

The wavelength of the audio waveform (λ) is the distance that the sound travels in a single direction along a medium in a repeating pattern between consecutive points of the same phase as observed in Fig. 1.2. The typical audio signal contains many different wavelengths with distinct amplitudes, but is typically represented by its Fourier transform as a sum of sinusoidal waves at different frequencies. The frequency of a signal, expressed in cycles per second, is expressed as,

$$f = \frac{c}{\lambda} \tag{1.1}$$

where c is the velocity of sound in air at room temperature with value of 347.87 m/s. Other values of sound velocity in other medium are shown in Table 1.1.

Table 1.1 Velocity of sound in different medium.

Medium	Velocity (m/s)
Rubber	40-150
Water	1433
Concrete	3200-3600
Bone	3750
Wood	4176
Glass, Pyrex	5640
Steel	5790

Frequency of a sound wave can be defined only if the wave is periodic in time. The fundamental frequency of the audio waveform is the greatest common divisor of the frequency of all the different frequency components of the signal. The typical audio frequency spectrum that is perceptible by humans ranges from 20 Hz to 20 kHz, or wavelengths from 16.5 mm to 16.5 m. However, most of the signal power in human speech is in the frequency band from 200 Hz to 4 kHz. The frequency band of a telephone is normally bounded from 300 Hz to 3kHz. Only the labial sounds and the fricative sounds have frequencies as high as 8 kHz to 10 kHz; however, there is relatively little power at these frequencies [3].

The high end of the audio frequency spectrum (5 kHz up to 20 kHz) is rarely processed in audio tracks, and it is only used in high definition audio applications like orchestra music reproduction. The infrasonic band is the band below the lowest frequency that can be heard, and these sounds are typically felt and not heard since some of organs of the human body can exhibit resonance at these frequencies. On the other hand, the ultrasonic band is the band above the highest frequency in the audible band, and these frequencies are used in ultrasonic cleaners, traffic detection, ultrasonic imaging in medical applications, burglar alarm systems, and remote controls. It is interesting to note that the hearing range for dogs (64 Hz to 44 kHz) includes ultrasonic frequencies.

The audio power amplifier in a sound system must be able to supply the high peak currents required to drive the , which is usually a low impedance load in the 4–8 Ω range. An ideal audio power amplifier must have zero output impedance to provide all the available power from the supply; however, the practical audio amplifier has a non-zero output impedance that must be small when compared to the loudspeaker impedance [3]. To understand the tradeoffs involved in the design of audio amplifiers, it is useful to review their history, the loudspeaker basics, and the metrics used to qualify their performance.

1.3 A brief history of audio amplifiers

Audio power amplifiers arose with the need of dealing with impulses which had to remain in a very definite time pattern to be useful. One of the earliest amplifying devices was the pipe organ. However, in a more generally accepted sense, amplifiers were invented when the nineteenth century technology became concerned with the transmission and reproduction of vibratory power: first sound, and then radio waves [4].

In 1876, Edison patented a device which he called the *aerophone*. It was a pneumatic amplifier in which the speaker's voice controlled the instantaneous flow of compressed air by means of a sound-actuated valve. The air was released in vibratory bursts similar to those that came from the speaker's mouth but more powerful and louder. Later on, an improved *aerophone* was attached to the phonograph [4].

The device which really opened up the field of amplification was the vacuum tube or valve. Early developments of the vacuum tube were pushed by Lee De Forest in 1906 with the introduction of the *audion* [5, 6]. The *audion* was a three terminal device partially enclosed in a glass tube that allowed the movement of electrons between two terminals, filament and plate, controlled by a third terminal called the grid. A small signal applied to the grid could control a large amount of current from the filament to the plate, allowing the amplification of electrical signals. An early application of vacuum tube was to transmit and receive radio waves. Other applications followed quickly: recording and reproduction of sound, detection and measurement of light, sound, pressure, etc. Thus, in early 1920s until 1950s the audio amplifiers used vacuum tubes such as the triode.

After the invention and refinement of the transistor in the early 1950s, the vacuum tubes were almost totally abandoned, and audio analog circuits became dominated by bipolar transistors using linear amplifiers. The transistor replaced the vacuum tube due to better reliability, cheaper cost, smaller size, and no warm up period [4]. Push-pull class-A amplifiers were dominant until the early 1960s. Designs using germanium devices appeared first, but suffered from the vulnerability of germanium to moderately high temperatures. After some time, all silicon transistors were developed using an NPN structure. However, most amplifiers using these transistors relied on input and output transformers for push-pull operation of the output stage. These transformers were heavy, bulky, expensive, and very nonlinear. Complementary bipolar power devices appeared in the late 1960s, allowing the development of full complementary output stages with less distortion than their predecessors [1]. The main remaining issue with the linear amplifiers was the power dissipation in the output stage, requiring bulky and expensive heat sinks to manage high output power.

The concept of class-D amplifiers was proposed by Baxandall [7] in 1959 for applications in LC oscillator circuits. Class-D amplifiers use switching to achieve a very high power efficiency, more than 92% in modern designs. By allowing each output switch device to be either fully turn off or on, losses are significantly reduced. Class-D amplifiers provide great advantages as far as heat dissipation, space efficiency and energy efficiency are concern. Currently, switching amplifiers are present in all practical applications like portable handsets, automotive, avionics systems, etc. A determining factor for the evolution of this technology is the emerging markets such as mobile phones, multichannel AV receivers, LCD and LED TV, portable navigation, medical equipment such as hearing aids, and the demand of high fidelity automotive audio systems. These growing markets demand small form-factor to fit in small fashionable enclosures with high-efficiency for thermal management, and increased battery lifetime while operating in high ambient temperature conditions. The lower power dissipation of class-D audio systems allows less heat, reduces cost and circuit board space, and augments battery life in portable systems.

1.4 Sound pressure level

The loudness of the sound wave has been difficult to characterize since each individual perceives the sound pressure differently, depending on age,

Table 1.2 SPL example levels.

Example	SPL (dB) at 1 m
Rustling of leaves	20
Quiet room	40
Conversation	60
Road with busy traffic	80
Noisy factory	90
Construction truck	100
Jet engine	120
Threshold of pain	140

lifestyle, health, among other circumstances. Therefore, a more formal metric is used to define how strong a sound wave is by measuring the difference, in a given medium, between a reference pressure (P_{ref}) and the pressure in the sound wave (P_{wave}). The unit to measure pressure is defined as a pascal ($Pa = 1\ \frac{\mathrm{N}}{\mathrm{m}^2} = 1\ \frac{\mathrm{kg \cdot m}}{\mathrm{s}^2}$), where N is the Newton, m is the meter, kg is the kilogram, and s is the second. As the human ear can detect sounds with a wide range of amplitudes, the sound pressure is often measured using a logarithmic scale such as the decibel. Therefore, the sound pressure level (SPL) can be defined as,

$$SPL = 20 \log \frac{P_{wave}}{P_{ref}} \tag{1.2}$$

where $P_{ref} = 20\ \mu\mathrm{Pa}$ is typically used since it is considered the threshold of human hearing for the sound propagating through air. The SPL can be measured using an instrument called a sound level meter [3] that senses the changes in pressure using a calibrated microphone and interprets the pressure difference to give a readout in the selected range. High SPL extended exposure can deteriorate a person's hearing by damaging sensitive inner-ear organs.

The typical SPL for conversational speech at 1m is 60 dB, while for a rock concert at 1m of the speaker is 100 dB; in mobile devices, the SPL performance can range from 60 dB up to 120 dB at short distances [8–10]. Other SPL examples are tabulated in Table 1.2 for different scenarios.

The importance of the SPL is that it gives a metric to compare different audio transducers for different scenarios. The overall audio system loudness will depend on how much SPL the system can produce at a given distance.

1.5 Loudspeaker transducers

The audio reproduction function in mobile devices can be classified in two applications. First, small audio transducers are used in headphone applications where the sound wave only travels a few centimeters into the ear canal; these are used commonly for hands-free conversations and music listening. Second, moderate audio transducers are used as loudspeakers for video conferences, video games, and other applications where the sound wave has to travel a few meters.

The electric impedance of the speakers used in these applications greatly influences the design of the audio amplifier. Thus, understanding of their physical construction and operation is needed in order to analyze their limitations and tradeoffs.

1.5.1 *Electromagnetic speaker*

The preferred speaker is the electromagnetic (EM) speaker, consisting of a magnet, a voice coil, and an acoustic cavity, as shown in Fig. 1.3. However, a large form factor is required in the EM speaker to deliver high SPL [11]. The typical materials used for the EM speaker construction are copper for the voice coil, plastic for the diaphragm, acrylonitrile butadiene styrene (ABS) for the frame, and Neodymium for the magnet. It operates by applying an electrical current through the voice coil to induce an electromagnetic field which in turn will generate a displacement of the acoustic diaphragm. Since the electromagnetic coupling factor is very small between the amount of electric current consumed to the amount of magnetic field produced, a large magnet and wide air cavity are needed to produce sound [3, 12].

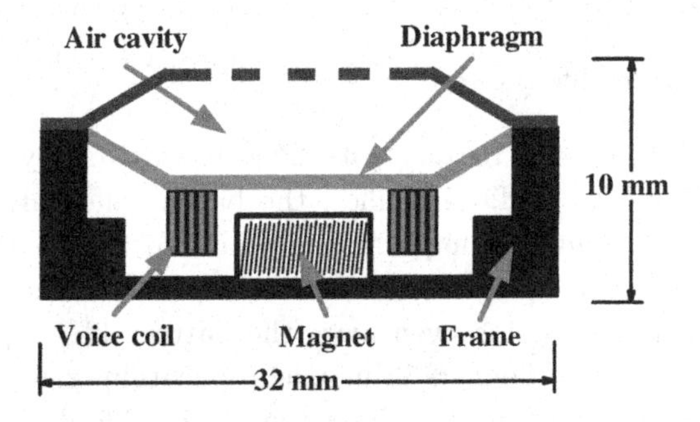

Fig. 1.3 EM speaker physical structure side view.

The electrical impedance of a typical EM speaker across the audio frequency bandwidth is shown in Fig. 1.4. On average, it behaves as a low value impedance between 4 to 32 Ω. A typical impedance value for most EM loudspeakers is 8 Ω while for EM headphones is 32 Ω. This means, that the amplifier has to output large electrical current through the voice coil to generate high SPL.

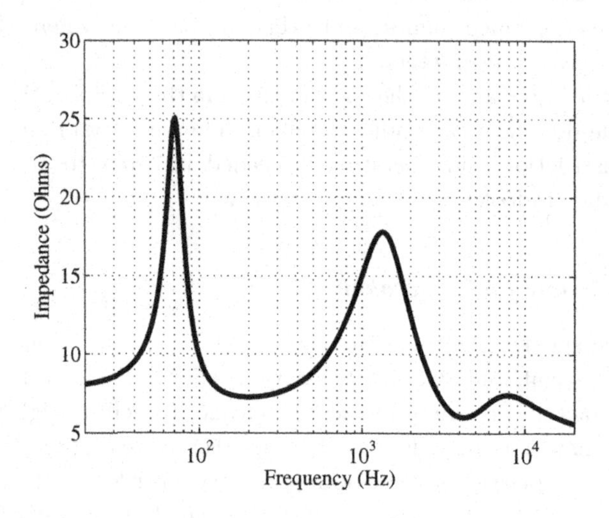

Fig. 1.4 EM electrical impedance versus frequency.

An interesting point to note is that the EM speaker's low impedance requires large output power to operate, quickly consuming the battery life of mobile devices. For example, to produce 90 dB of SPL from an 8 Ω EM speaker, the battery has to provide around 1 W of average power or 353 mA of load current [11, 13, 14]. The battery life can be calculated as,

$$\text{Battery life (hours)} = \frac{\text{Battery capacity (mAh)}}{\text{Load current (mA)}}. \tag{1.3}$$

If a typical lithium-ion battery with 2000 mAh capacity is used with an ideal 100% efficient audio amplifier, the battery life only considering the EM speaker current consumption would be 5.67 hours. In real applications, the audio amplifier current consumption would also be included in the calculation, decreasing even more the battery life; typical current consumption for audio amplifiers in mobile applications range from 1 mA to 10 mA. Thus, EM loudspeakers limit the battery life despite the audio amplifier's high efficiency and low current consumption.

1.5.2 *Piezoelectric speaker*

The physical structure of a typical piezoelectric (PZ) speaker is shown in Fig. 1.5 where a PZ element is attached to a film encased between a front panel and rear panel. Typical materials used for its construction are polycarbonate for the front and real panel, plastic resin or metal for the film, and lead zirconate titanate for the PZ element. The PZ element deflects with voltage applied across its terminals, causing the film to warp and bend up and down according to the voltage applied across the PZ element. The deflecting/bending action creates pressure waves pushing air through one or more openings that are arranged on the front panel that resonate and amplify the response of the speaker.

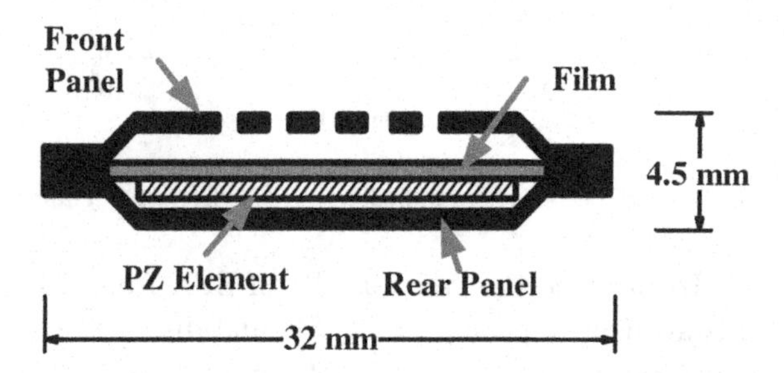

Fig. 1.5 PZ speaker physical structure side view.

The PZ element in the speaker is typically a multilayer ceramic component that behaves electrically as a capacitor across the audio frequency bandwidth [15]. Figure 1.6 shows the measured impedance versus frequency of a typical PZ speaker in comparison to the EM speaker impedance. It can be observed that for most of the audio frequency spectrum, the PZ speaker has an impedance orders of magnitude larger than the EM speaker impedance. This allows the audio amplifier to use very low power to operate the speaker, improving the battery life of mobile devices.

The capacitive behavior of the PZ speaker is highly reactive, meaning that the energy applied to the transducer is stored and most of it is returned to the supply each signal cycle. Ideally, this will allow almost no average power consumption from the battery, but the PZ speaker has some dielectric losses in the ceramic material that will dissipate some power as heat. Typical dissipation factors range from 0.4% up to 1% with quality factors > 50 for most PZ speakers available.

The following example will illustrate the PZ speaker's low power

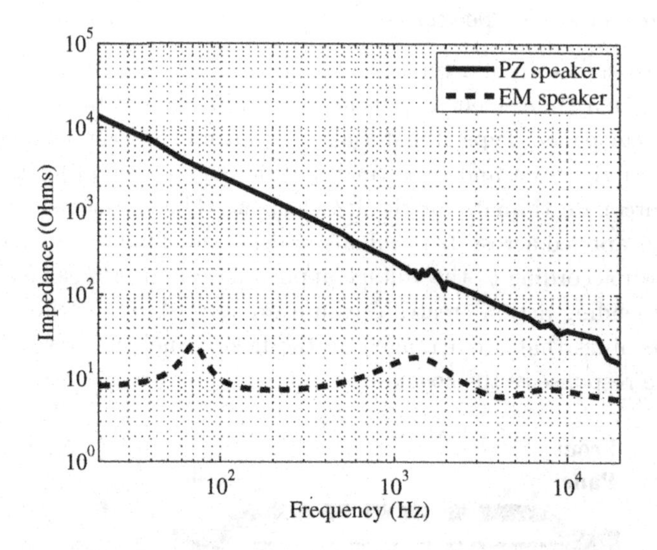

Fig. 1.6 PZ and EM speakers impedance versus frequency comparison.

consumption. To produce the same 90 dB of SPL from the PZ speaker with a 338 Ω equivalent impedance at 1 kHz and dissipation factor of 1%, the battery only needs to provide 1.2 mW of average power or 133 μA of load current. If the same lithium-ion battery with 2000 mA/h capacity is used solely for the audio amplifier driving the PZ speaker, the battery life calculated using Eq. (1.3) would be 15,000 hours. That is a battery life extension of 2645x times compared to the EM speaker. However, in real applications, the audio amplifier current consumption would be dominant, limiting the battery life. If the audio amplifier has a current consumption of 2.46 mA, the real battery life would be 770 hrs which is considerably large compared to the EM speaker example. The PZ speaker provides low power consumption and high SPL, making it an attractive alternative for mobile devices, especially when used with low power high efficiency audio amplifiers.

Typical voltage levels across the PZ speaker terminals should be in the range of 10-20 V_{pp} to achieve the maximum SPL, and could be generated from the battery using high-efficiency step-up voltage circuits [16–18]. Commercial audio amplifiers for PZ speakers provide high-voltage outputs using these circuits, but their distortion and power consumption is still large [19–22]. Thus, new circuits for PZ speakers that dissipate less power and produce less distortion are desirable.

Another area of interest for PZ transducers in mobile devices is for haptic feedback systems, where a PZ actuator (similar in structure to a PZ speaker) is driven by a power amplifier to create mechanical vibrations felt by the user [23]. The power amplifier for haptic feedback systems delivers power to the load at the desired mechanical resonant frequency to provide the maximum displacement of the PZ actuator. Thus, power amplifiers for PZ actuators in mobile devices with low power dissipation will always be required to allow small form factors and long battery life.

1.6 Performance metrics of audio amplifiers

To compare the quality of different audio amplifiers, it is necessary to understand the performance metrics in the audio amplification process. The metrics can be divided into two main groups based on the measurements required to obtain them: (1) frequency measurements, and (2) power measurements. The former includes total harmonic distortion (THD), total harmonic distortion plus noise (THD+N), signal-to-noise ratio (SNR), power-supply rejection ratio (PSRR), and power-supply induced intermodulation distortion (PS-IMD). The latter encloses the power efficiency (η).

The basic equipment for CDA measurements must include: an audio analyzer or spectrum analyzer, oscilloscope, highly-linear signal generator, evaluation board (printed circuit board) with device under test (DUT), multimeter, power supply, power resistors, and low-pass filter components [24–26].

A general test setup for frequency measurements is shown in Fig. 1.7. It includes a System One Dual Domain Audio Precision (AP) [25, 26] that generates the highly-linear test signal (V_{IN}), a bias and power network for the DUT from the power supply, the low-pass external LC filter, and the speaker load (Z_L). The Audio Precision equipment also provides a built-in spectrum analyzer to obtain the frequency response of the output signal.

The speaker load (Z_L) is typically replaced with an emulated load with equivalent impedance to control the accuracy of the test; for EM speakers the emulated load is a power resistor (4-32 Ω) in series with an inductor which value varies depending on the speaker power rating, and for PZ speakers the emulated load is a high-voltage ceramic capacitor which value depends on the emulated PZ speaker voltage rating.

The most used signal in the measurements is a sine wave, typically at 1 kHz. This is because a sine wave corresponds to a single tone in the frequency domain, and the non-linear output signal can be expressed as a collection of sine waves where the more dominant frequency harmonics are

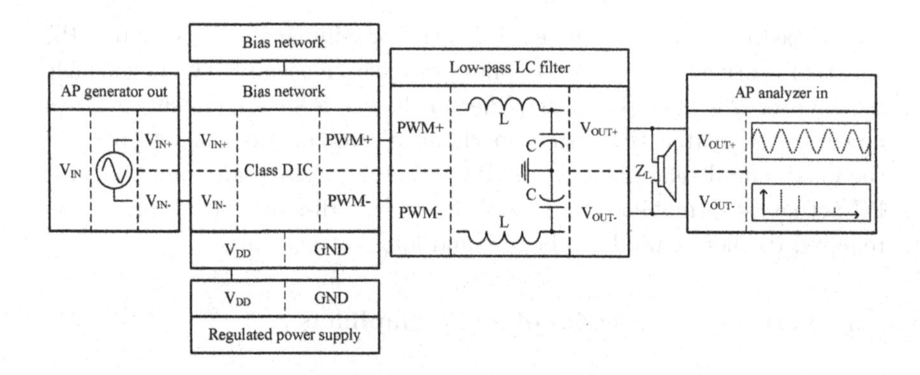

Fig. 1.7 General test setup for CDA frequency measurements.

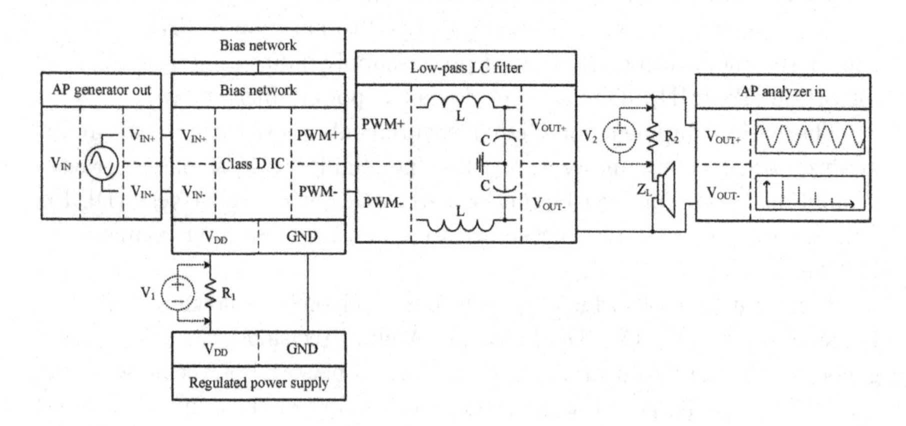

Fig. 1.8 General test setup for CDA power measurements.

within the audio frequency band. Thus, they are easy to identify and use for various performance metrics.

The power measurements are taken with the general test bench shown in Fig. 1.8. The addition of power resistors R_1 and R_2 allow the measurement of the input current flowing from the power supply V_{DD}, and the current through the load Z_L, respectively, with the aid of multimeters V_1 and V_2. The power resistors have very small values (<1 Ω) to avoid high power dissipation in them.

1.6.1 *Total harmonic distortion plus noise*

The total harmonic distortion plus noise (THD+N) metric measures the amount of distortion that is generated by the amplification process compared with the fundamental input frequency, including the total noise produced by the amplifier. The THD+N is defined as the ratio of the fundamental frequency power to the sum of the harmonics power plus noise power as,

$$\mathrm{THD+N} = \sqrt{\sum_{i=2}^{N} \frac{V_i^2}{V_1} + \frac{V_n^2}{V_1}} \tag{1.4}$$

where V_i is the RMS voltage of the nth harmonic, V_n is the integrated noise RMS voltage in the bandwidth of interest, and V_1 is the fundamental frequency RMS voltage. If the noise is not accounted and only the linearity is of interest, then a total harmonic distortion (THD) calculation can be used as,

$$\mathrm{THD} = \sqrt{\sum_{i=2}^{N} \frac{V_i^2}{V_1}}. \tag{1.5}$$

The THD+N metric is the most accepted definition for audio quality of the system since real amplifiers will have noise that is not accounted in the THD metric. For example, for different amplifiers that have the same THD but they operate at different power levels, a large V_1 would reduce the effect of V_n in the THD+N; but for a small V_1, the contribution of V_n is larger, increasing the THD+N, as observed in Eq. (1.4). Thus, using the THD metric to compare amplifiers with different output power levels would be unfair.

The THD+N is typically measured against a sweep of output amplitudes for a single signal frequency, or for a sweep of frequencies from 20 Hz to 20 kHz for a single output amplitude. Since the THD+N varies several orders of magnitudes across the whole output power range, the measured value can be expressed in logarithmic scale or as a percentage. Some typical value are summarized in Table 1.3, and can be calculated as,

$$\mathrm{THD+N\,(dB)} = 20 \cdot \log \frac{\mathrm{THD+N\,(\%)}}{100}. \tag{1.6}$$

That is,

$$\mathrm{THD+N\,(\%)} = 100 \cdot 10^{\left(\frac{\mathrm{THD+N\,(dB)}}{20}\right)}. \tag{1.7}$$

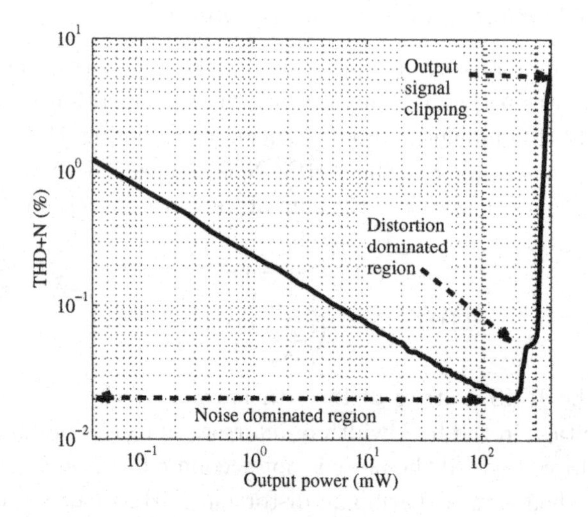

Fig. 1.9 THD+N typical plot against output power for audio amplifiers.

Table 1.3 THD+N measurement units.

Decibel (dB)	Percentage (%)
0	100
−20	10
−40	1
−60	0.1
−80	0.01
−100	0.001

A typical THD+N plot against the output power for a 1 kHz signal is illustrated in Fig. 1.9, where 3 regions can be identified. The first region is at low output powers where the THD+N value is dominated by the noise of the circuit. The second region is at medium to high output power where the THD+N value is dominated by the distortion of the output signal. The third region is when the output signal amplitude reaches the supply voltage value and it starts to clip, increasing the distortion drastically.

Other frequency tones can be used to compare the THD+N of an audio amplifier for a fixed output amplitude. A special case is for a 6.6 kHz input signal since the output THD+N would be dominated by the third harmonic (V_3=19.8 kHz) that is at the high limit of the audio frequency spectrum. This test signal gives the worst case scenario THD+N number of the amplifier. Typical THD+N values for commercial audio amplifiers in mobile devices range from −65 dB to −110 dB.

1.6.2 *Signal-to-noise ratio*

The signal to noise ratio (SNR) is a metric that defines the ratio of the signal power to the noise power of the amplifier. Noise sources come from the power supply hum (60 Hz), switching noise, thermal noise from the circuit components in the amplifier, and radio frequency interference. This noise could be audible and will limit the smallest signal that can be processed by the amplifier. The SNR can be expressed as,

$$\mathrm{SNR} = 10\log\frac{P_o}{P_n} \cong 20\log\frac{V_o}{V_n} = 20\log V_o - 20\log V_n. \tag{1.8}$$

where V_o is the output voltage amplitude of the fundamental frequency of the audio signal, and V_n is the integrated output noise of the amplifier from 20 Hz to 20 kHz. Since the human hearing has limited capabilities at very low and very high frequencies outside the audible band, certain weighting factors, such as the A-weighting, are used to filter out the undesired frequencies to define a measurement bandwidth for the noise [27].

One method to calculate the SNR is by using the frequency spectrum representation of the output signal with noise, and use the noise floor to extract V_n. The noise floor is obtained by measuring the output of the audio amplifier with a very small or no audio signal.

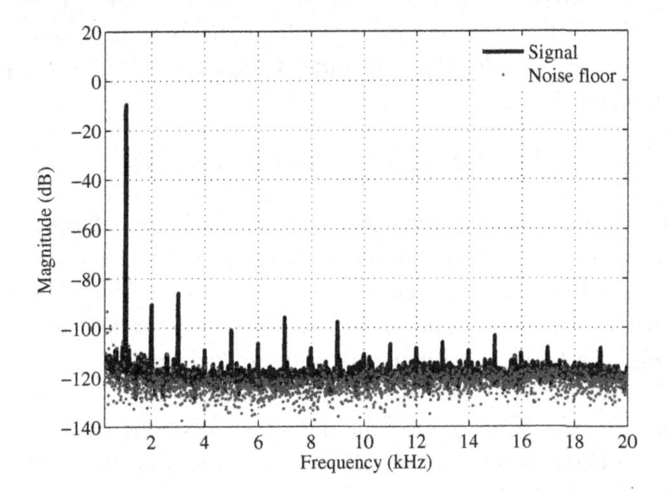

Fig. 1.10 Typical output frequency spectrum for audio amplifiers.

Figure 1.10 shows a 4096-point frequency spectrum from a fast-Fourier transform (FFT) for an audio amplifier where the noise floor and the harmonics for a 1 kHz signal can be observed. Note that the fundamental tone is −12 dB, and the noise floor is around −120 dB across the audio bandwidth. From this plot, we can assume that $V_n(dB)$ is −120 dB but this would be incorrect since the FFT effectively creates a discrete representation of the analyzed signal, and the number of discrete points taken during the FFT calculation distributes the signal energy in a finite number of frequency bins. This FFT processing gain has the effect of lowering the energy content on each frequency bin as the number of points is increased, and has to be corrected when calculating V_n from the FFT. For this example the noise floor is around −120 dB, but the 4096-point FFT is effectively lowering the noise floor level by around $10 \log(4096/2) \cong 33$ dB. Thus, the noise floor level is at −87 dB, and the resulting SNR is $-12 - (-87) = 75$ dB.

1.6.3 *Power supply rejection ratio*

The power supply rejection ratio (PSRR) is a metric that defines the ratio between the output signal to the noise signal introduced by the supply of the amplifier. This metric is important because battery-powered devices share the audio amplifier's power supply plane with the same noisy power supply plane of digital circuits. The supply noise mixes with the audio and carrier signals, degrading the overall THD+N performance. The supply noise rejection needs to be high over the whole audio frequency spectrum, to avoid a degradation of the THD+N. Since the supply noise could be orders of magnitude smaller than the output signal, the PSRR is typically expressed in decibels as,

$$\text{PSRR} = 20 \log \frac{V_n}{V_o} = 20 \log V_n - 20 \log V_o. \tag{1.9}$$

This metric is measured when there is no audio signal present (idle condition), and the noise signal present in the supply voltage node is swept across the audio frequency spectrum. The test setup for PSRR is shown in Fig. 1.11 where the noise signal is superimposed on the DC supply level with the help of a high-power linear amplifier or driver.

A special case for the PSRR is with the noise signal at 217 Hz; this tone is especially important for audio amplifiers in cell phone devices since it represents the GSM burst used for the device communication. Therefore, if the audio amplifier is intended for a cell phone application, it must have a high PSRR performance also at low frequencies.

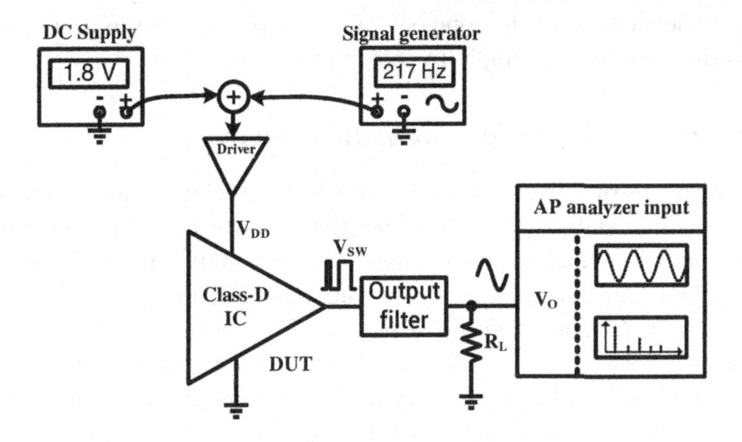

Fig. 1.11 Test setup for PSRR measurement of single-ended CDA.

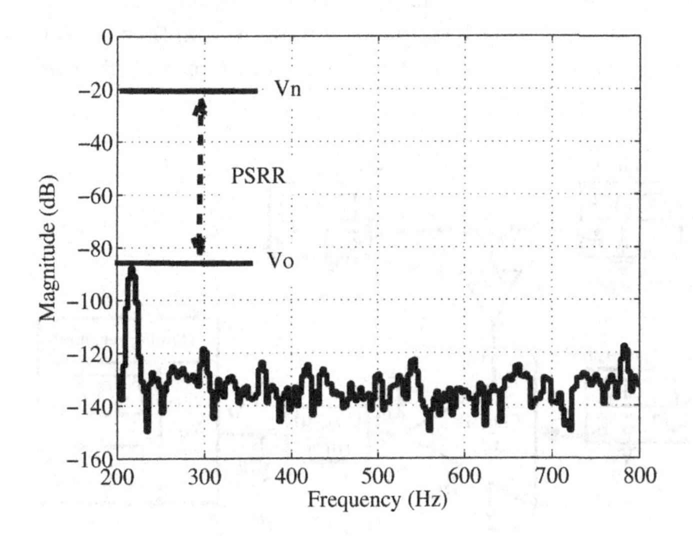

Fig. 1.12 PSRR measurement example for a noise signal at 217 Hz with −20 dB amplitude.

Figure 1.12 illustrates the output signal frequency spectrum for an amplifier in idle state (no audio signal) with a −20 dB tone signal at 217 Hz at its supply voltage node. Since the plot shows the values in logarithmic scale, the PSRR is the difference between the V_o and V_n in decibels that for this example is around 65 dB. The PSRR performance of the audio amplifier is highly dependent on the topology of the amplifier and the output stage connecting the battery supply to the output.

1.6.4 *Power supply intermodulation distortion*

The PSRR metric characterizes the supply noise rejection at the idle condition when no audio signal is present. However, during normal operation, the supply noise and the audio signal are present in the amplifier. The power supply intermodulation distortion (PS-IMD) measures the interaction between the noise and audio signals. The PS-IMD is the amplitude modulation between noise and audio signals at different frequencies. The intermodulation products occur since all amplifiers are non-linear circuits that generate harmonics, and they are located at multiples of the sum and difference frequencies of the audio and noise signal frequencies.

The test setup to measure the PS-IMD is shown in Fig. 1.13 where a 1 kHz sine wave is used as the input of the audio amplifier with a superimposed noise signal with 0.1 V_{pp} at 217 Hz at the amplifier supply source. This test setup is very similar to the PSRR setup but the difference is that an input signal is present.

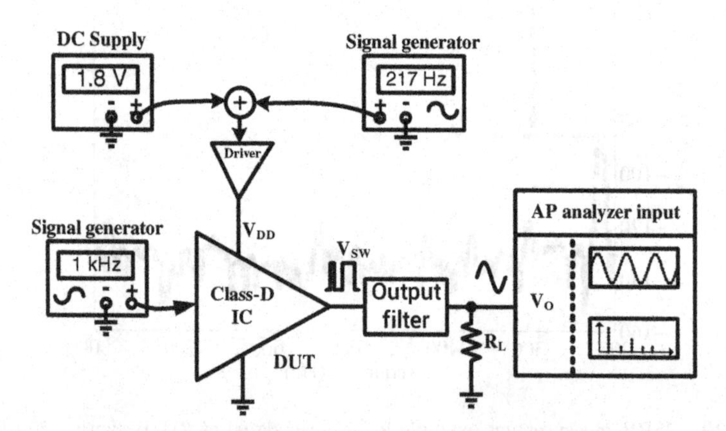

Fig. 1.13 Test setup for CDA PS-IMD measurement.

The PS-IMD can be measured from the frequency spectrum of the output signal, as show in Fig. 1.14. An audio signal at 1 kHz is applied to the amplifier, and a supply noise signal at 217 Hz is superimposed on the supply voltage node. These two signals will generate two dominant intermodulation products (i.e. 1kHz $\pm$ 217Hz) at 783 Hz and 1217 Hz. The PS-IMD can

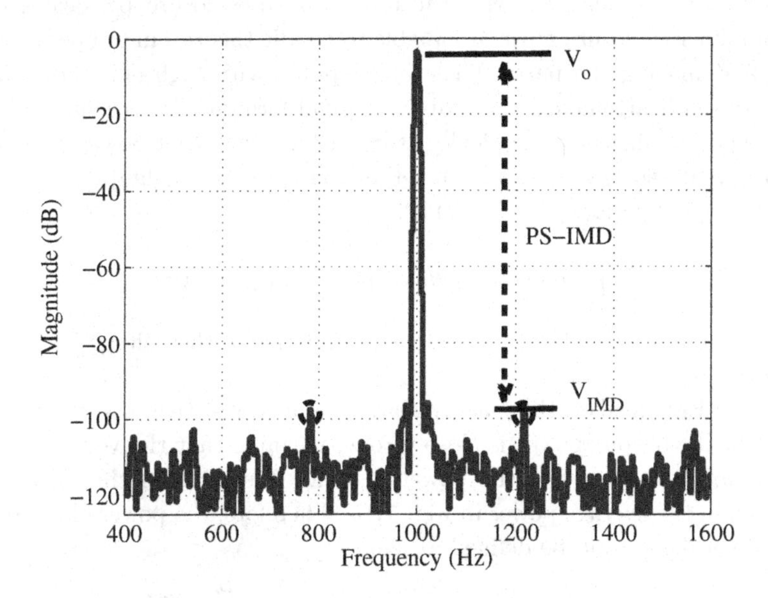

Fig. 1.14 PS-IMD measurement example for a noise signal at 217 Hz and audio signal at 1 kHz.

be expressed as the magnitude ratio between the intermodulation products to the fundamental as,

$$\text{PS-IMD} = 20\log\frac{V_o}{V_{IMD}} = 20\log V_o - 20\log V_{IMD} \tag{1.10}$$

where V_{IMD} is the intermodulation product amplitude, and V_o is the fundamental audio signal amplitude. For this example, the PS-IMD would be $-6-(-98) = 92$ dB. The PS-IMD metric is highly correlated with the linearity of the system and its PSRR performance. If the amplifier has poor PSRR and high distortion, then the PS-IMD will be poor too.

1.6.5 *Power efficiency*

One of the most important metrics for audio amplifiers is the power efficiency (η). This metric represents how much of the energy provided by the battery is effectively used for the intended purpose of audio amplification. In other words, if the audio amplifier consumes/dissipates power due to its biasing or power loss, then the efficiency will be less than 100%. The power dissipated by the amplifier as heat will limit the maximum output power that it can deliver to the load for a defined enclosure/package size. The amplifier enclosure must be capable to handle this rise in temperature without damaging the part. Thus, an amplifier with high efficiency will dissipate less heat, making it suitable for small-form factor enclosures.

The power efficiency is typically estimated as the output power divided by the input (supply) power, using the average power definition over a sinewave signal period ($T = 2\pi/\omega$) as,

$$P_{avg} = \frac{1}{T}\int_0^T v(t)\cdot i(t)\cdot dt = V_{RMS}\cdot I_{RMS}\cdot\cos(\varphi) \tag{1.11}$$

where the phase angle between the current relative to the voltage is represented by φ.

For an EM speaker, the load appears as almost resistive, and the term $\cos(\varphi)$ in the output power is close to one, meaning that the voltage and current are in phase, and the power is being dissipated in the load, as observed in the average power in Fig. 1.15. Therefore, the power efficiency for EM speakers could be defined as [28],

$$\eta = \frac{P_{o,avg}}{P_{i,avg}} = \frac{P_{o,avg}}{P_{o,avg}+P_{loss,avg}} = 1 - \frac{P_{loss,avg}}{P_{i,avg}} \tag{1.12}$$

where the $P_{o,avg}$ is the average output power delivered to the load, $P_{i,avg}$ is the average input power consumed from the battery, and $P_{loss,avg}$ is the average power loss in the audio amplifier. In general, most of the audio amplifier applications are targeted for EM speakers; thus, this efficiency definition is typically used.

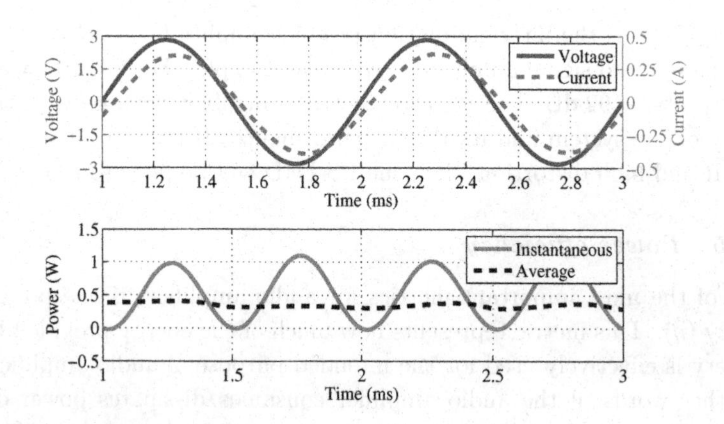

Fig. 1.15 Instantaneous and average power for a EM speaker with $\varphi < 15°$.

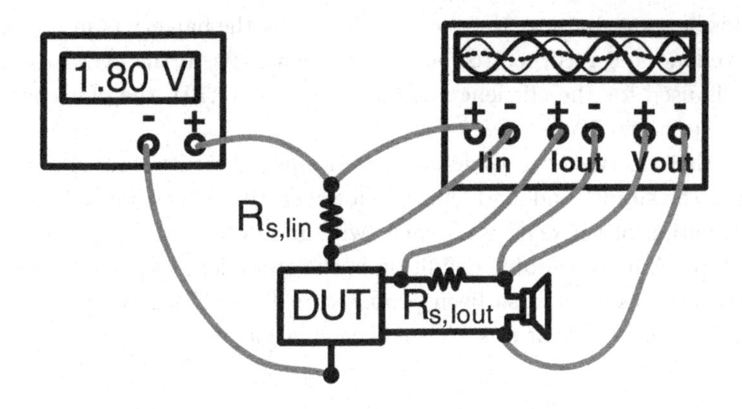

Fig. 1.16 Test setup for CDA efficiency measurement.

The test setup for efficiency measurement is shown in Fig. 1.16 where power resistors $R_{s,Iin}$ and $R_{s,Iout}$ are used to measure the average input and output power to calculate the efficiency.

The PZ speaker is a highly reactive speaker alternative that offers very low power consumption. Its capacitive nature needs a different definition of power efficiency since the current leads the voltage by almost 90 degrees, causing the term $\cos(\varphi)$ to be close to zero, and appearing as if very little power is being dissipated by the load, as observed in the average power in Fig. 1.17.

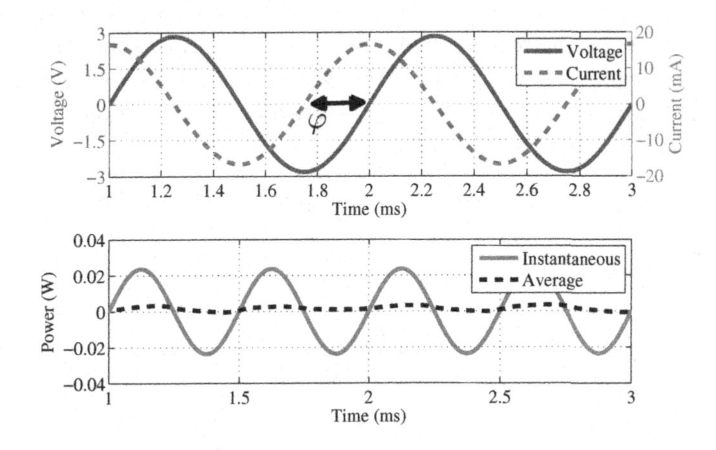

Fig. 1.17 Instantaneous and average power for a PZ speaker with $\varphi \cong 90°$.

This happens because the energy supplied by the battery is stored in the reactive load and returned to the battery each cycle. If the average output power is used for the efficiency definition in Eq. (1.12), the efficiency will appear very low [29].

Another alternative is to define the efficiency in terms of energy transfer between the supply and load [30,31]. However, the energy analysis requires an estimation of the energy for each switching cycle, making it a complex procedure. A more suitable definition of the amplifier's power efficiency for capacitive transducers has been proposed in [32–34], where the apparent power (P_{app}) is used for the efficiency calculation.

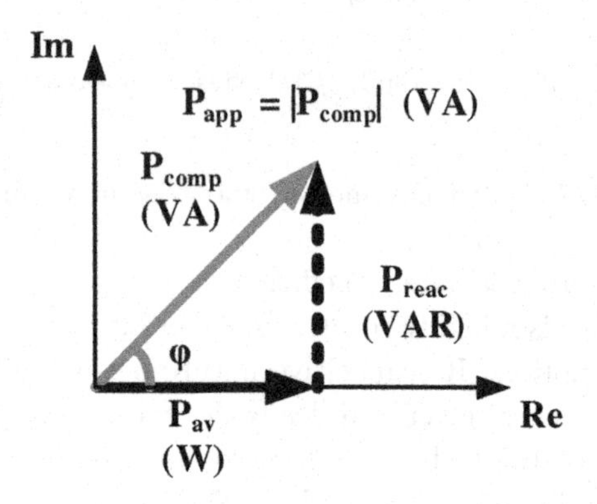

Fig. 1.18 Complex power definition.

The apparent power, measured as $P_{app} = V_{RMS} \cdot I_{RMS}$, is the magnitude of the complex power vector which contains the information of the reactive power (P_{reac}) and the average or real power (P_{av}), as observed in Fig. 1.18. The complex power is the general representation for the voltage and current product, and it can be expressed as,

$$\vec{P}_{comp} = \vec{P}_{av} + \vec{P}_{reac} = \vec{P}_{comp} \cos\varphi + \vec{P}_{comp} \sin\varphi. \tag{1.13}$$

Thus, the amplifier's power efficiency for capacitive loads is defined as [32–34],

$$\eta_{PZ} = \frac{P_{o,app}}{P_{i,app}} = \frac{P_{o,app}}{P_{o,app} + P_{loss,app}} = \frac{1}{1 - \frac{P_{loss,app}}{P_{o,app}}}, \tag{1.14}$$

$$P_{o,app} = V_{o,RMS} \cdot I_{o,RMS} \cong \frac{V_{o,RMS}^2}{|Z_L(j\omega)|} \tag{1.15}$$

where $Z_L(j\omega)$ is the equivalent impedance of the PZ speaker at the operating frequency. This efficiency definition states that the amplifier has to process the apparent power required by the PZ speaker with the minimum power dissipation. In other words, the average input power reflects the power dissipation of the system. This is important since the audio amplifier has to be designed for a large capacitive load with low power dissipation. This definition is useful for all PZ actuators in general.

1.7 Audio amplifier classification

The audio amplifier can process the audio signal as a linear operation continuously in time and amplitude, or as a non-linear operation continuously in time but discretely in amplitude. A linear audio amplifier processes the signal with an output stage configured as a current source, while a non-linear audio amplifier has a switching output stage that processes the signal with a modulation scheme.

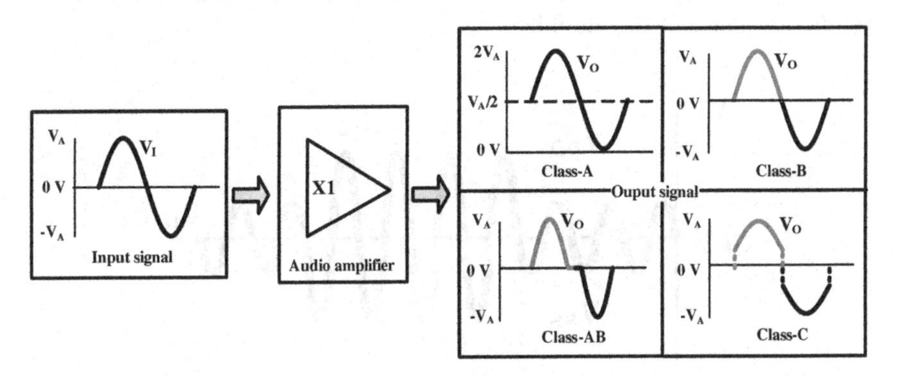

Fig. 1.19 Main linear amplification classes.

The main linear amplification classes are summarized in Fig. 1.19. The audio amplifier is typically configured as a voltage follower with enough output current to drive the impedance of the speaker. The main amplifier linear classes are the class-A, class-B, class-AB, and class-C. The linear amplifiers' biasing point defines the range of the input signal that they can amplify at the output with a tradeoff between linearity and power dissipation.

The class-A amplifier outputs 100% of the signal swing, providing minimum distortion; but, its output stage biasing point is placed at half the maximum amplitude of the signal; hence, dissipating large power.

The class-B amplifier's biasing point is chosen to output only 50% of the signal swing; the small biasing point decreases the power consumption, but at the expense of large distortion. The class-AB operation combines the reduced power dissipation and low distortion of the class-B and class-A, respectively; but, it requires a complex biasing scheme to operate [35, 36]. The class-C amplifier's biasing point is chosen to output less than 50% of the output swing to lower the power dissipation drastically; however, its high distortion prohibits its use for audio applications.

Other more advanced amplifier classes have been proposed to increase the power efficiency of linear amplifiers such as the class-G and class-H. The class-G amplifier provides better power efficiency compared with the class-AB operation [37–40]. The efficiency improvement is achieved by reducing the supply voltage for smaller output signals, and thus, reducing the power dissipation. The power-supply transition is achieved without affecting the dynamic range of the output signal, as observed in Fig. 1.20. However, the distortion during the supply transition can be detrimental in the performance of the amplifier.

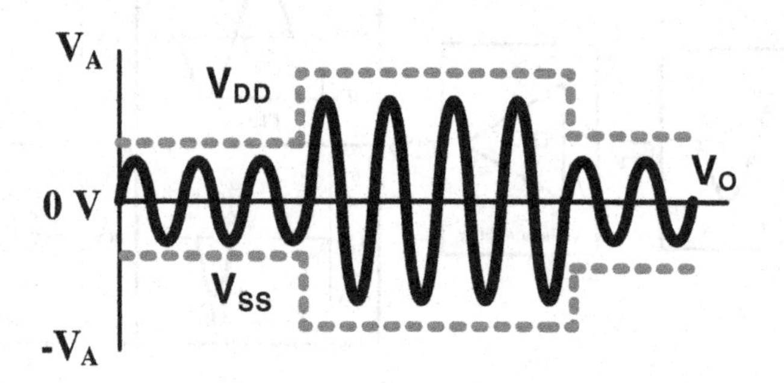

Fig. 1.20 Class-G amplifier operation.

The class-H amplifier operates in a similar way as the class-G amplifier by continuously changing the supply voltages as observed in Fig. 1.21. The smooth supply transitions allow very low distortion and improved efficiency in the amplification process since the supply is high only when needed by the audio signal. However, a dedicated power management circuit is required to adapt the supply voltages according to the input signal, degrading the overall efficiency benefits, and increasing the cost and power consumption of the audio amplifier.

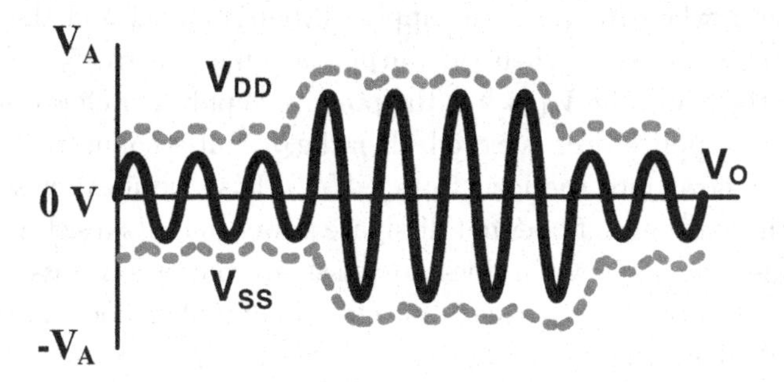

Fig. 1.21 Class-H amplifier operation.

The class-G and class-H amplifiers improve the power efficiency of the amplifier by switching the supply voltage for different levels of output signal. However, the biasing's power dissipation still exist during the amplification process. Another alternative is to keep a fixed supply voltage but switch the output signal between the supply voltages to operate the output stage as a digital switch. This operation is known as a class-D operation, where the continuous input signal is modulated by a high frequency carrier that generates a stream of pulses that are applied to the load, as observed in Fig. 1.22.

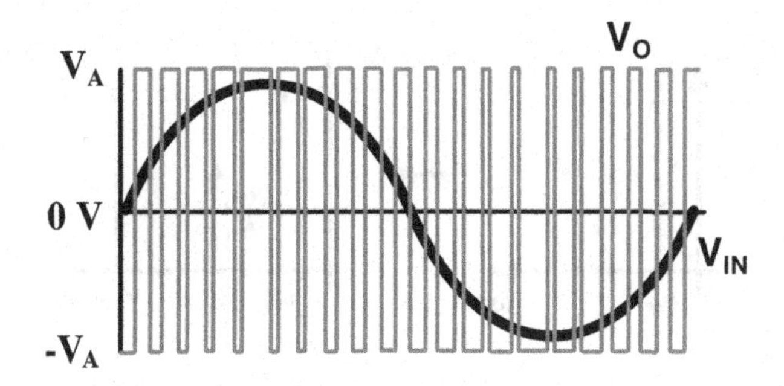

Fig. 1.22 Class-D amplifier operation, time domain.

The advantage of this discontinuous operation of the class-D amplifier compared to the conventional amplifier classes is that ideally there is no power dissipation in the amplifier. The class-D amplifier is comprised by two switches that connect the output to either the positive supply voltage (V_A) or to the negative supply voltage ($-V_A$). The class-D basic operation is as follows: when the positive supply switch is closed and the negative supply switch is opened, then the output is connected to V_A and current flows into the load from V_A; when the positive supply switch is opened and the negative supply switch is closed, then the output is connected to ground and current flows into the load from $-V_A$. Thus, the ideal closed switch carrying the load current does not dissipate power since its resistance is zero, and the ideal opened switch does not dissipate power since its resistance is infinite. In other words, the ideal class-D amplifier does not dissipate power while delivering all the available power to the load.

The ideal efficiency for a class-D amplifier can be 100%, meaning that all the power from the supply is delivered to the speaker. In reality, the maximum efficiency is limited by the finite switch resistance in the output stage, the output filter component power loss, the amplifier quiescent power, and the output stage capacitive power loss. The class-D amplifier is the focus of this book due to its high power efficiency.

The class-D operation can be also understood by looking at the output frequency spectrum of the modulated signal, as shown in Fig. 1.23. The frequency spectrum has the audio fundamental frequency (ω_o) and harmonics ($n\omega_o$) plus the modulation carrier signal frequency (ω_{SW}) located at higher frequencies out of the audio band. The harmonics are generated since the non-ideal modulator and limited slew rate in the output stage distort the signal.

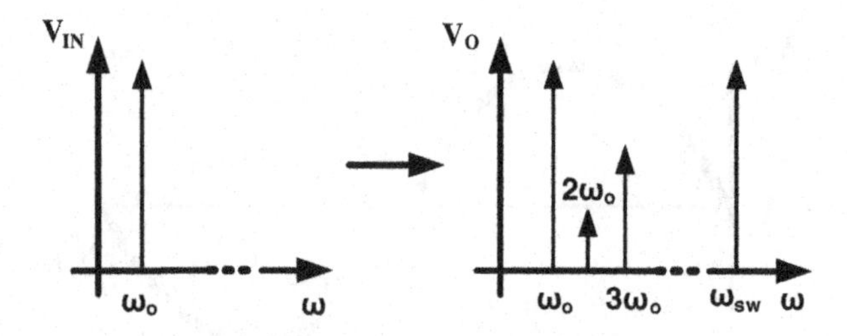

Fig. 1.23 Class-D amplifier operation, frequency domain.

The low frequency components of the modulated output signal represents the desired audio information. Thus, a passive low pass filter is used to recover the audio signal information at the speaker. This output filter is typically implemented with an inductor and capacitor to avoid degrading the efficiency. The cut-off frequency for the output filter is usually set to 20 kHz to include the whole audio frequency band.

Chapter 2

Principles of Class-D Audio Amplifiers

2.1 Class-D amplification

The class-D amplifier (CDA), also known as digital power amplifier or switching amplifier, is an electronic device which takes an input voltage signal, either in analog or digital domain, and amplifies it using an output stage operating as a digital inverter. The main advantages of the class-D amplification are its high efficiency and its robust digital output signal.

The class-D output stage operation is as follows. If the input signal is in analog domain, the audio signal is typically modulated by a high-frequency carrier signal to obtain a pulse-width modulation (PWM) and then it is amplified by the class-D output stage. Figure 2.1 shows a single-ended open loop CDA for analog inputs, where the high-frequency carrier signal (V_C) is used to achieve the PWM of a low-frequency input signal (V_I).

If the input signal is in digital domain expressed as a bit stream of $n = 8$ up to $n = 48$ bits, the input digital signal vector ($D_I\ [0 : (n-1)]$) is typically transformed from a pulse-code modulation (PCM) to PWM using a digital block that perform some signal processing and data rate reduction using the data clock signal (D_{CLK}), as observed in Fig. 2.2. The PCM uses a unique digital code while the PWM uses a unique pulse width to represent an output voltage level, as illustrated in Fig. 2.3. Typical PCM digital vectors are of 16 bits or ($D_I\ [0:15]$), requiring a complex and precise digital circuit to translate the information to a PWM signal.

In both cases, the modulated signal (V_{PWM}) is used to switch the output power transistors between the voltage rails with high efficiency. Finally, an output low-pass filter is used to recover the low-frequency signal and apply it to the speaker.

Typical operating frequencies for the CDA in audio applications are

20 Hz–20 kHz for the input and output signals, and 200–400 kHz for the carrier and PWM signals.

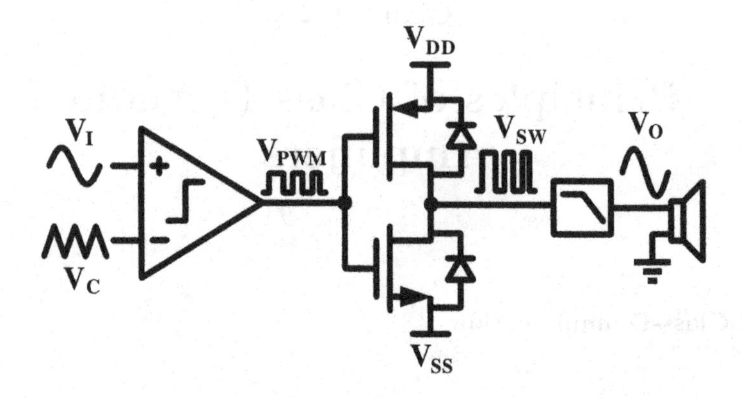

Fig. 2.1 Analog class-D amplifier in single-ended open-loop architecture.

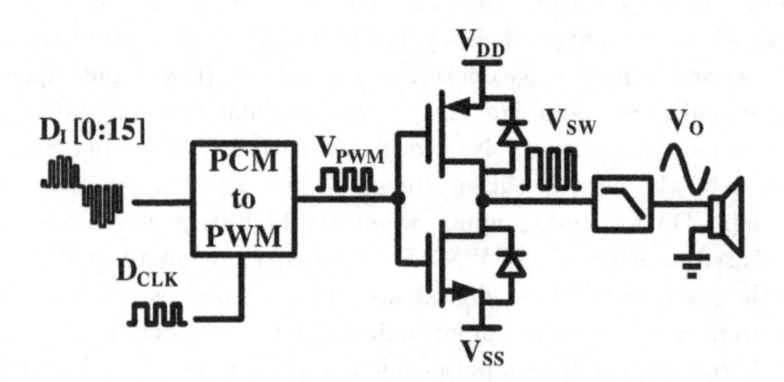

Fig. 2.2 Digital class-D amplifier in single-ended open-loop architecture.

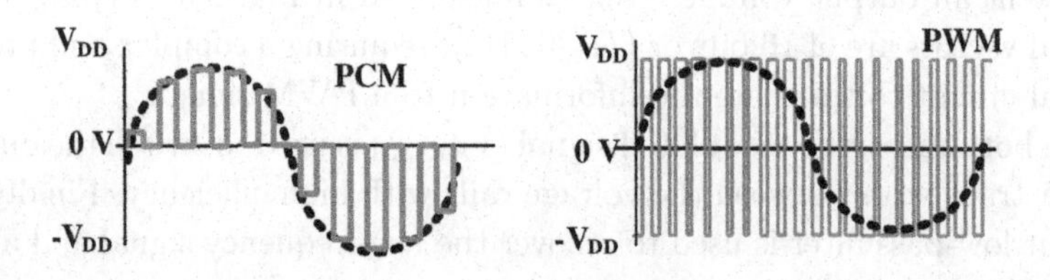

Fig. 2.3 PCM and PWM comparison.

The audio information in some devices is processed in digital domain, making it more convenient to use the digital information as the input of the CDA. However, the digital input CDA is typically used in open loop architectures, requiring complex signal processing and calibration algorithms to achieve high performance. On the other hand, the analog input CDA can be used with a feedback mechanism that helps to correct distortion, noise, and enhance the audio performance in general, allowing low power operation.

The only drawback for analog input CDA is the need for a digital-to-analog converter (DAC) to transform the digital audio signal to analog domain in order to amplify it. The analog input CDA is the preferred choice for class-D amplification in mobile devices due to its low power operation, and is the focus of this book. Thus, all the following discussion is referred to analog input class-D audio amplifiers. It is important to notice that the output stage of the CDA, as shown in Fig. 2.1, is very similar to the output stage of a step-down DC-DC Buck converter [41–43]. The main difference is that the goal in a Buck converter is to regulate the output voltage as a constant DC voltage source under different load conditions, where the voltage V_I is a constant voltage reference, and the duty cycle of V_{PWM} is proportional to the desired output voltage. Also, the output filter is designed to reduce the voltage ripple at the output. Thus, all the modulation and design techniques discussed in this book could be extended to the DC-DC Buck converter.

2.2 Advantages and disadvantages of class-D amplifiers

The main advantage of the class-D amplification is its high efficiency and low power dissipation that allows extended battery life in mobile devices. The output stage in CDAs is typically implemented using CMOS transistors operating as switches. When the switch is open, they appear as very high resistor ($> 1M\Omega$), having an ideal zero power dissipation since no current is flowing through them; when the switch is closed, they appear as very low resistor ($< 0.1\Omega$), having an ideal zero power dissipation since there is no voltage drop across them. This operation allows low power dissipation in the switch. Thus, the heat sink typically used in other amplifier classes can be drastically reduced or completely removed.

One of the disadvantages of the class-D amplification is that its output signal is a square wave at full power that needs to be removed before applying it to the speakers. This requires an output filter with external components that occupy PCB area and increase the bill-of-materials. However, alternative techniques can be used to minimize the output filter requirements with switching strategies that provide multi-level output signals [44,45].

Another disadvantage for the CDA is the electromagnetic interference (EMI) radiated by the inductance of the cables and/or PCB traces connecting the CDA with the speaker [46]. This is particularly important in mobile devices since most of the circuits are placed closely. Thus, sensitive analog circuits such as analog-to-digital converters (ADC), radio frequency receivers, and voltage or current references can be drastically affected by the EMI. Several techniques to improve the EMI can be used to spread the energy of the high-frequency carrier signal used in PWM such as spread spectrum or edge-rate control, at the expense of additional power consumption and design complexity [47,48].

2.3 Class-D output stage power losses

The ideal CDA can reach 100% efficiency. However, the CDA power losses due to its implementation will limit the maximum efficiency. A comprehensive analysis for the power losses in switching power stages can be found extensively in the literature [28,49–52]. The power efficiency in the class-D amplifier is defined as,

$$\eta \cong \frac{P_o}{P_o + P_{loss}}, \tag{2.1}$$

$$P_{loss} \cong P_Q + P_{CL} + P_{SW} + P_{BD} \tag{2.2}$$

where the power losses in the CDA (P_{loss}) is mainly dominated by the amplifier quiescent power (P_Q), the conduction losses (P_{CL}), switching losses (P_{SW}) and body-diode losses (P_{BD}) of the output stage.

Conduction losses occur due to the ohmic losses of the output switches' drain to source ON resistance (R_{dsON}) and are more prominent when the current demanded by the load is large. Switching losses occur due to the power dissipated in the switch during the transition from ON to OFF states, and by the charging and discharging of parasitic capacitors, especially in the output stage. Body-diode losses occur due to the body-diode conduction and its reverse recovery charge that could be considerable for large output currents. The body-diode is inherent in the power device with its source tied to its bulk forming a PN junction from the bulk to the drain as illustrated in Fig. 2.1. The power losses can be expressed as,

$$P_{CL} \cong I_{o,RMS}^2 \cdot R_{dsON} \tag{2.3}$$

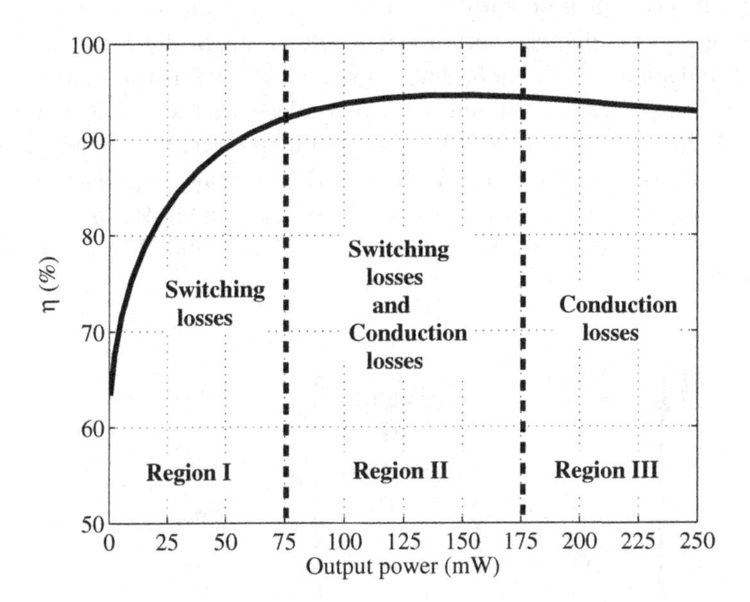

Fig. 2.4 Efficiency versus output power example for a class-D amplifier, showing power losses dominated region.

$$P_{SW} \cong \sum_i \left(F_{SW} \cdot V_{CP}^2 \cdot C_{P,i}\right) + V_{DD} \cdot I_{o,PK} \frac{2 \cdot t_{trans}}{t_{SW}}, \tag{2.4}$$

$$P_{BD} \cong V_{SD} \cdot F_{SW} \cdot (I_{o,PK} \cdot t_{deadtime} + I_{rrm} \cdot t_{rr}) \tag{2.5}$$

where $I_{o,RMS}$ is the output RMS current, F_{SW} is the CDA switching frequency, $C_{P,i}$ are the parasitic capacitors in the output stage, V_{CP} is the voltage across each $C_{P,i}$, V_{DD} is the supply voltage, t_{trans} is the transition time from ON to OFF of the switch, t_{SW} is the switching period, V_{SD} is the body-diode source-to-drain voltage, $I_{o,PK}$ is the peak output current, $t_{deadtime}$ is the deadtime used to avoid shoot-through current, I_{rrm} is the body-diode maximum reverse recovery current, and t_{rr} is the body-diode reverse recovery time. Deadtime or non-overlap using shoot-through current [53]. The CDA efficiency can be characterized in three regions across its operating output power, as shown in Fig. 2.4. It can be observed that at low power levels (region I), the P_{loss} is dominated by P_{SW} and P_Q; at medium power levels (region II), all the components of P_{loss} contribute to the total; and, at high power levels (region III), the P_{loss} is dominated by P_{BD} and P_{CL} since the output current is large.

The output stage is designed to minimize the power losses in a particular operating region; different optimizations will result in different efficiency curves, as depicted in Fig. 2.5. The peak in each curve is the result of the optimized output stage to minimize the power losses in the region of interest for the desired application. The main goal in high power CDA applications is to improve the efficiency in region II and III since they require low power dissipation for reduced size and weight. The main goal in low power CDA applications is to improve the efficiency in regions I and II to extend the battery life.

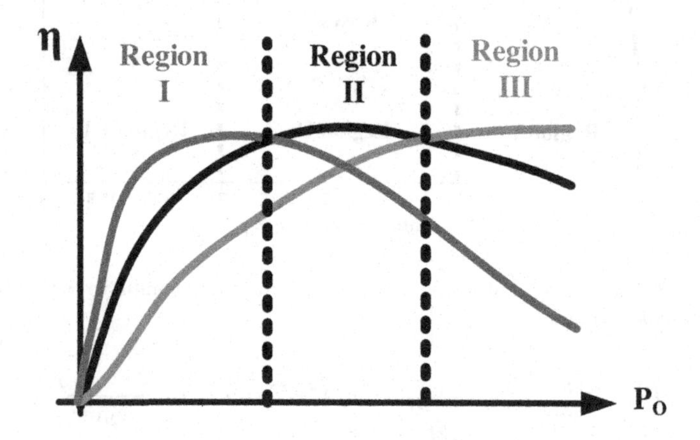

Fig. 2.5 Efficiency curves for different output stage optimizations.

Different audio transducers will influence the impact of each power loss on the efficiency. Figures 2.6 and 2.7 illustrate the output stage of the CDA when driving the electrical models of an electromagnetic (EM) speaker or a piezoelectric (PZ) speaker, respectively. The main contributors of P_{SW} are the input and output capacitance ($C_{P,i}$) of the output stage that can be large if the switches are sized to obtain small R_{dsON}. However, a large transistor switch will have a large body-diode, increasing the contribution of P_{BD} to the total power losses, and limiting the maximum efficiency.

The advantage of using the PZ speaker is that its high impedance requires small current to operate, minimizing the impact of P_{CL} in the efficiency. This would allow smaller output switches to obtain the same P_{CL} but will decrease the P_{SW}, enhancing the overall efficiency. Moreover, the small output current together with a short $t_{deadtime}$ will reduce the P_{BD} contribution to the total power losses.

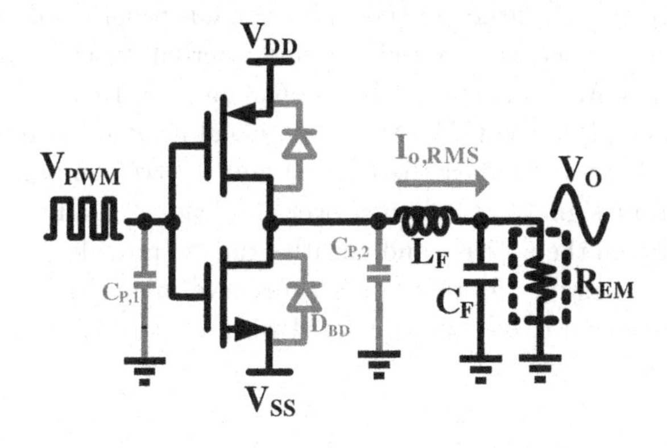

Fig. 2.6 Class-D output stage driving an EM speaker load.

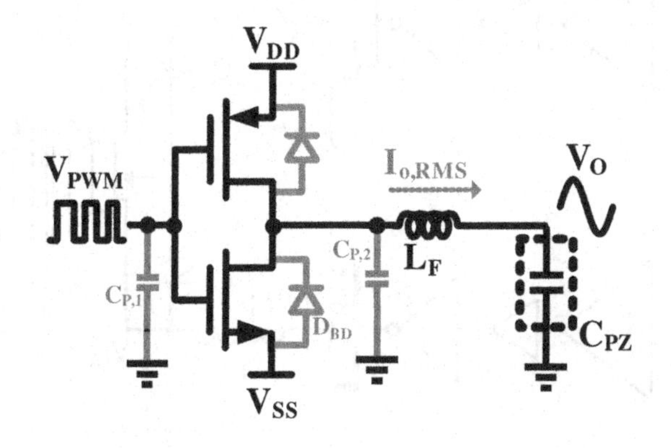

Fig. 2.7 Class-D output stage driving a PZ speaker load.

2.4 Open loop class-D amplifiers with pulse-width modulation

Class-D audio amplifier based on pulse-width modulation (PWM), or naturally sampled pulse-width modulation, scheme is the most used architecture [2]. Its circuitry is simple, stable and with low-power consumption. The typical architecture of a class-D audio power amplifier based on PWM have been shown in Fig. 2.1.

Pulse-width modulation is based on the simple fact that the mean value of a two-level square wave is proportional to its duty cycle. The modulation is done by comparing the signal to a constant slope carrier. However, the audio signal must be bounded to avoid clipping in the modulation process.

Consequently, the amplitude of the audio signal is usually normalized with respect to the carrier wave amplitude and referred to as the modulation index, M where $M \in [0, 1]$ [54]; for example, an $M = 1$ corresponds to the carrier peak amplitude equal to the input signal peak amplitude.

The switching method of the class-D audio power amplifier describes how the output signal is controlled over the load. Usually, the signal is switched between the positive and negative rails to provide a binary signal. However, it is possible to generate a three level modulated signal under certain conditions if the amplifier is used with a bridged configuration.

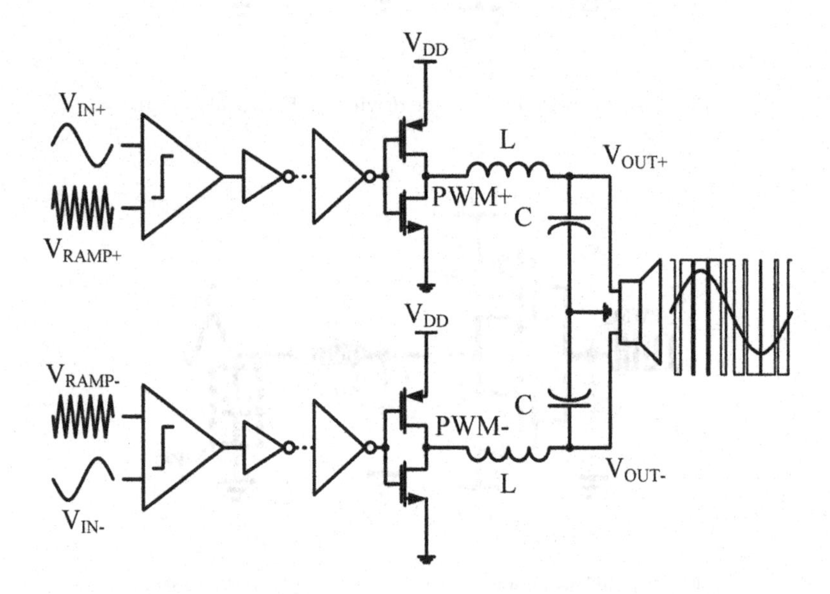

Fig. 2.8 Two-level bridge-tied-load class-D audio power amplifier.

A class-D audio power amplifier with two-level bridge-tied-load (BTL) topology is shown in Fig. 2.8. As it can be appreciated, the amplifier is the fully-differential version of the conventional class-D amplifier in Fig. 2.1. This modulation is also known as pulse-width modulation AD, or PWM AD [2]. It produces a binary signal with minimum cross-over distortion and zero common-mode components. On absence of signal at the input, the binary signal is a square wave with a 50% duty cycle. Many commercial products are based on this modulation scheme.

Figure 2.9 illustrates the generation of the two-level pulse-width modulated signal. The audio signal and the carrier wave generate two

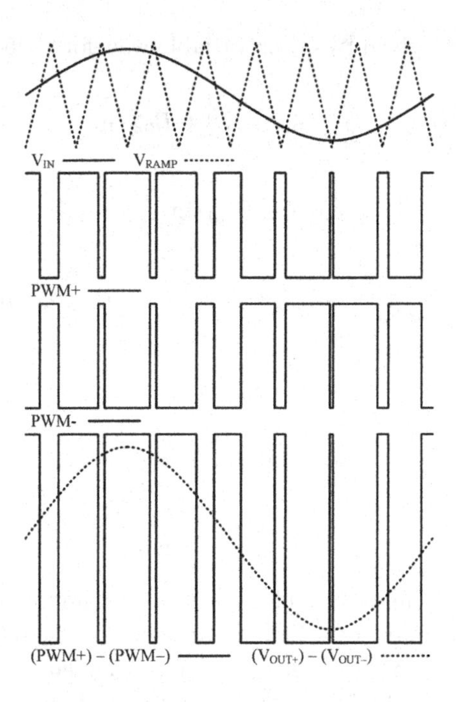

Fig. 2.9 Generation of two-level pulse-width modulated signal.

complementary fully-differential digital signals (PWM+ and PWM−) whose differential voltage is doubled.

Determination of the harmonic frequency components of a pulse-width modulated signal requires using a fast Fourier transform (FFT) analysis of a simulated time-varying switched waveform. This approach offers the benefits of a reduced mathematical effort but requires considerable computing capacity and usually leaves uncertainty to simulation errors (specially in systems with higher carrier frequencies) due to time resolution of the simulation and the periodicity of the overall waveform. In contrast, an analytical solution which exactly identifies the harmonic components of a pulse-width modulated signal ensures that the correct harmonics are being considered. The most well-known analytical method of determining the harmonic components of a pulse-width modulated signal is using the double Fourier integral analysis (DFIA) [54, 55]. This analysis assumes the existence of two time variables, which can be thought of as the high-frequency modulating wave (carrier signal) and the low-frequency modulated wave (baseband audio signal). In general, the value of the function $f(t)$, representing the modulation of a given two time variable waveforms [54], is given by

$$
\begin{aligned}
f(t) = \frac{A_{00}}{2} &+ \sum_{n=1}^{\infty} [A_{0n} \cos\beta + B_{0n} \sin\beta] \\
&+ \sum_{m=1}^{\infty} [A_{m0} \cos\alpha + B_{m0} \sin\alpha] \\
&+ \sum_{m=1}^{\infty} \sum_{\substack{n=-\infty \\ (n \neq 0)}}^{\infty} [A_{mn} \cos(\alpha + \beta) + B_{mn} \sin(\alpha + \beta)] \qquad (2.6)
\end{aligned}
$$

where

$$\alpha = m(\omega_c t + \theta_c), \qquad (2.7)$$

$$\beta = n(\omega_o t + \theta_o) \qquad (2.8)$$

and m is the carrier index variable, n is the baseband index variable, ω_c is the carrier angular frequency, θ_c is an arbitrary phase offset angle for the carrier waveform, ω_o is the baseband angular frequency, θ_o is an arbitrary phase offset angle for the baseband waveform, and A_{mn} and B_{mn} are the coefficients of the magnitudes of the harmonic components. In general, the two angular frequencies (ω_c and ω_o) will not be an integer ratio.

The first term of Eq. (2.6), $A_{00}/2$ where $m = n = 0$, corresponds to the DC offset component of the pulse-width modulated waveform. The first summation term, where $m = 0$, defines the output fundamental low-frequency (audio signal) and its baseband harmonics (if any). This term includes low-order undesirable harmonics around the fundamental output (harmonic distortion) which should be minimized. The second summation term, where $n = 0$, corresponds to the carrier wave harmonics, which are relatively high frequency components, since the lowest frequency term is the modulating carrier frequency. The final double summation term, were $m,\ n \neq 0$, is the combination of all possible frequencies harmonics formed by taking the sum and difference between the modulating carrier waveform harmonics and the reference waveform and its associated baseband harmonics. These combinations are usually called sideband harmonics [54].

The pulse-width modulated waveform in the class-D audio power amplifier can be generated using different carrier waveforms. The most popular are sawtooth waveform (Fig. 2.10), triangle waveform (Fig. 2.13), and exponential-shaped waveform (Fig. 2.22).

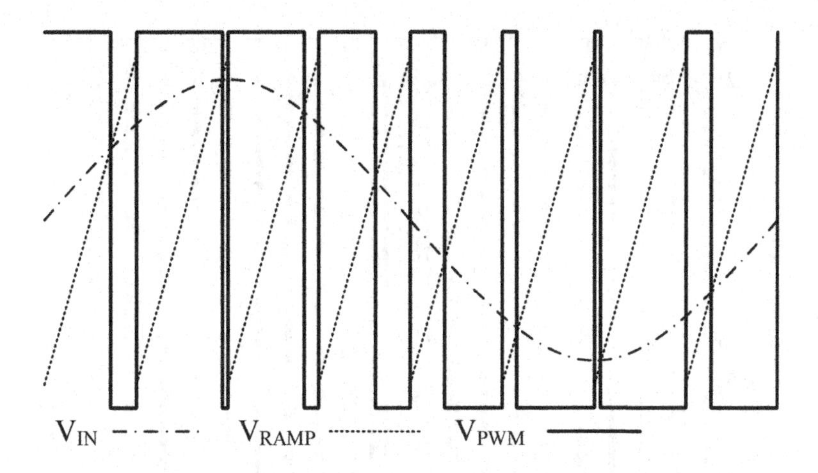

Fig. 2.10 Sine-sawtooth pulse-width modulation.

Figure 2.10 illustrates an example of naturally sampled modulation using a sawtooth waveform, also called *trailing-edge naturally sampled modulation (TENSM)* [54], where V_{IN}, V_{RAMP}, and V_{PWM} are the input audio signal, the sawtooth carrier waveform and the pulse-width modulated signal. Hence, considering a sinusoidal input audio signal V_{IN} of the form

$$V_{IN}(t) = M\cos(\omega_o t + \theta_o) \tag{2.9}$$

where M is the modulation index, ω_o is the target output frequency, and θ_o is an arbitrary output phase. Then, the resulting pulse-width modulated (V_{PWM}) signal using a sawtooth waveform (V_{RAMP}) can be expressed in terms of its harmonics components, by using the double Fourier integral analysis [54], as

$$\begin{aligned} V_{PWM}(t) &= V_{DC} + V_{DC}M\cos(\omega_o t + \theta_o) \\ &+\frac{2}{\pi}V_{DC}\sum_{m=1}^{\infty}\frac{1}{m}[\cos(m\pi) - J_0(m\pi M)\sin\alpha] \\ &+\frac{2}{\pi}V_{DC}\sum_{m=1}^{\infty}\sum_{\substack{n=-\infty \\ (n\neq 0)}}^{\infty}\frac{1}{m}J_n(m\pi M)\begin{bmatrix}\sin\left(n\frac{\pi}{2}\right)\cos(\alpha+\beta) \\ -\cos\left(n\frac{\pi}{2}\right)\sin(\alpha+\beta)\end{bmatrix} \end{aligned} \tag{2.10}$$

where V_{DC} is the DC offset component, $J_0(\cdot)$ and $J_n(\cdot)$ are the Bessel functions of the first kind [54], and α and β are the arguments defined in Eq. (2.7) and Eq. (2.8).

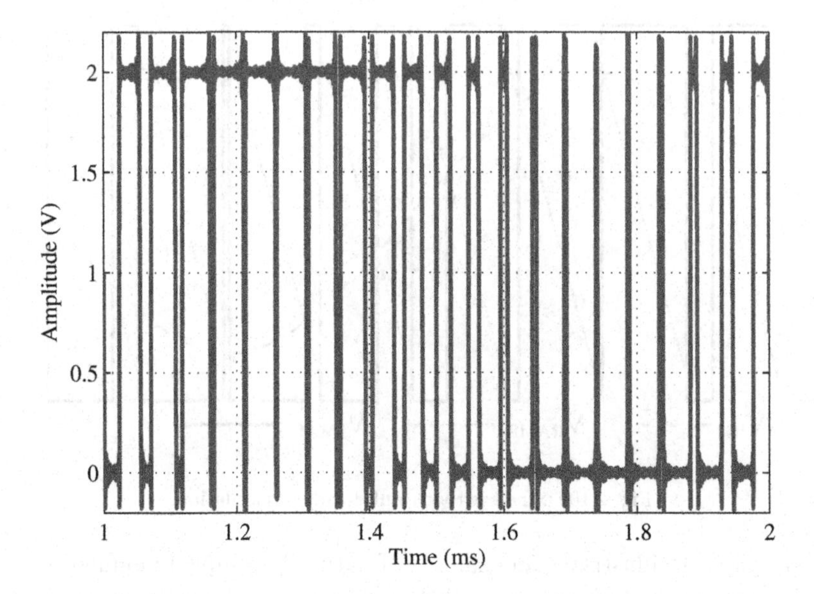

Fig. 2.11 Two-level pulse-width modulated signal with trailing-edge naturally sampled modulation.

The two-level pulse-width modulated signal with trailing-edge naturally sampled modulation in Eq. (2.10) is plotted in Fig. 2.11 with coefficients $m = n = 50$, modulation index $M = 0.9$, $V_{DC} = 1\,V$ and $\omega_c \,/\, \omega_o = 21$. Also, in Fig. 2.12, the spectrum of the pulse-width modulated signal with trailing-edge naturally sampled modulation is presented. Notice that there are no baseband harmonics in the pulse-width modulated signal, since the second term in Eq. (2.10) contains only the fundamental tone, therefore the harmonic distortion is zero.

The more common form of naturally sampled pulse-width modulation employs a triangular carrier wave instead of a sawtooth carrier as shown in Fig. 2.13. This type of modulation is also called *double-edge naturally sampled modulation* (*DENSM*) [54].

The pulse-width modulated signal using a triangular carrier waveform can also be expressed in terms of its harmonic components. Hence, by using the sinusoidal waveform V_{IN} given in Eq. (2.9), and applying the double Fourier integral analysis [54], the pulse-width modulated (V_{PWM}) signal

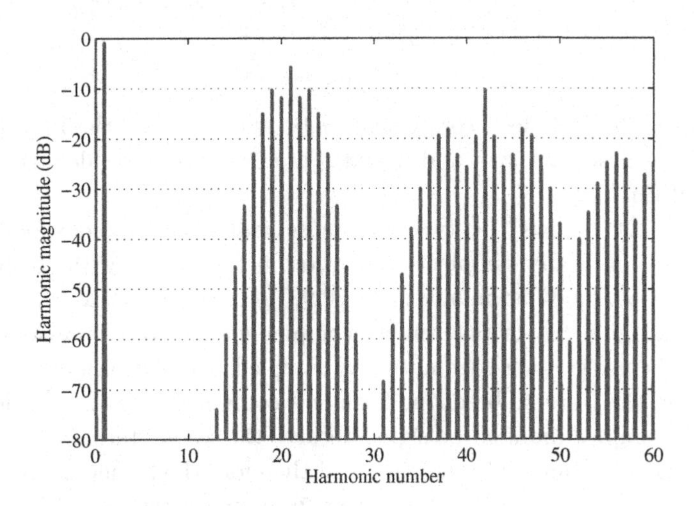

Fig. 2.12 Harmonic components of two-level pulse-width modulated signal with trailing-edge naturally sampled modulation.

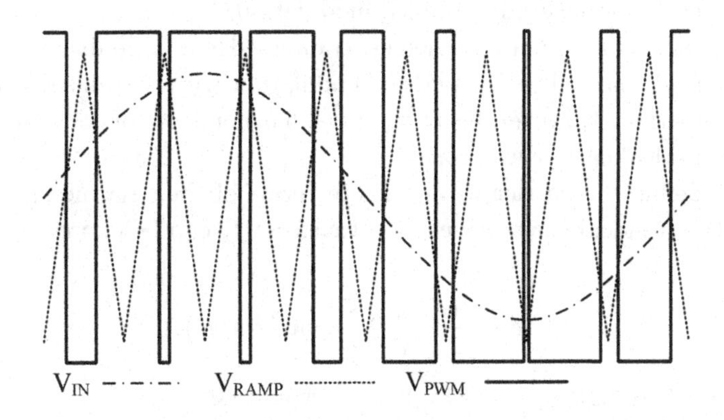

Fig. 2.13 Sine-triangle pulse-width modulation.

with double-edge naturally sampled modulation scheme is given by

$$\begin{aligned} V_{PWM}(t) &= V_{DC} + V_{DC} M \cos(\omega_o t + \theta_o) \\ &+ \frac{4}{\pi} V_{DC} \sum_{m=1}^{\infty} \frac{1}{m} J_0\left(m\frac{\pi}{2}M\right) \sin\left(m\frac{\pi}{2}\right) \cos\alpha \\ &+ \frac{4}{\pi} V_{DC} \sum_{m=1}^{\infty} \sum_{\substack{n=-\infty \\ (n\neq 0)}}^{\infty} \frac{1}{m} J_n\left(m\frac{\pi}{2}M\right) \\ &\times \sin\left([m + n]\frac{\pi}{2}\right) \cos\gamma \end{aligned} \tag{2.11}$$

where

$$\gamma = \alpha + \beta \tag{2.12}$$

and V_{DC} is the DC offset component, $J_0(\cdot)$ and $J_n(\cdot)$ are the Bessel functions of the first kind [54], and α and β are the arguments defined in Eq. (2.7) and Eq. (2.8).

Figure 2.14 shows the two-level pulse-width modulated signal with double-edge naturally sampled modulation in Eq. (2.11) with coefficients $m = n = 50$, modulation index $M = 0.9$, $V_{DC} = 1\,V$ and $\omega_c \,/\, \omega_o = 21$, and Fig. 2.15 illustrates its harmonic components. As in the previous modulation scheme (trailing-edge naturally sampled modulation), expressed in Eq. (2.10), there are no baseband harmonic components in the double-edge naturally sampled modulation signal because the second term in Eq. (2.11) contains the fundamental tone only, and therefore the harmonic distortion is also zero. Also notice that the power spectrum of the modulated signal with double-edge naturally sampled modulation contains much less carrier harmonic components than the power spectrum of the modulated signal with trailing-edge naturally sampled modulation.

The previous modulation schemes, trailing-edge and double-edge naturally sampled, have shown a perfect modulation with zero baseband harmonic distortion. However, in reality, total harmonic distortion is non-zero due to non-ideal carrier waveforms.

If we define the carrier waveforms in terms of their harmonic components, then an ideal sawtooth carrier waveform f_s can be expressed as

$$f_s(t) = \frac{1}{2} - \frac{1}{\pi}\sum_{i=1}^{\infty}\frac{1}{i}\sin\left(2i\pi t f_c\right) \tag{2.13}$$

and an ideal triangle carrier waveform f_t would be

$$f_t(t) = \frac{8}{\pi^2}\sum_{i=1,3,5,\ldots}^{\infty}\frac{(-1)^{(i\,-\,1)\,/\,2}}{i^2}\sin\left(2i\pi t f_c\right). \tag{2.14}$$

Figures 2.16 and 2.17 show the sawtooth carrier waveform and the triangle carrier waveform, respectively, in terms of their harmonic components, and Fig. 2.18 shows the carrier waveforms with only fifty harmonic considered (i = 50). Therefore, it would be necessary to implement an infinite bandwidth system to generate a perfect carrier signal to obtain zero harmonic distortion in the class-D audio power amplifier. Unfortunately, band-limited systems degrade the performance of the overall class-D audio amplifier by generating undesired baseband harmonic components.

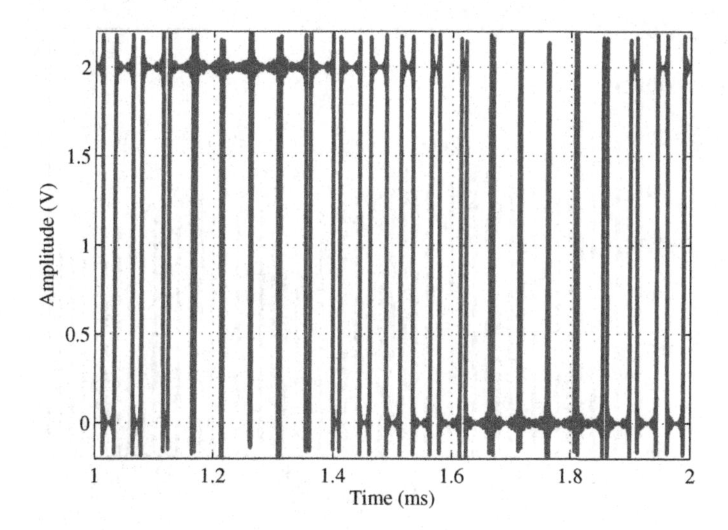

Fig. 2.14 Two-level pulse-width modulated signal with double-edge naturally sampled modulation.

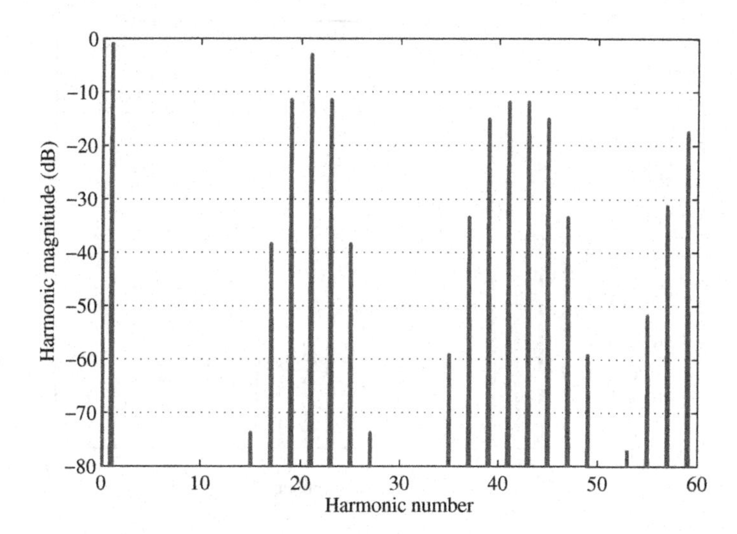

Fig. 2.15 Harmonic components of two-level pulse-width modulated signal with double-edge naturally sampled modulation.

It is possible to quantify the linearity performance of class-D amplifiers operating in open loop by analyzing the harmonic components of their carrier waveforms. A similar analysis has been proposed in [55] in order to

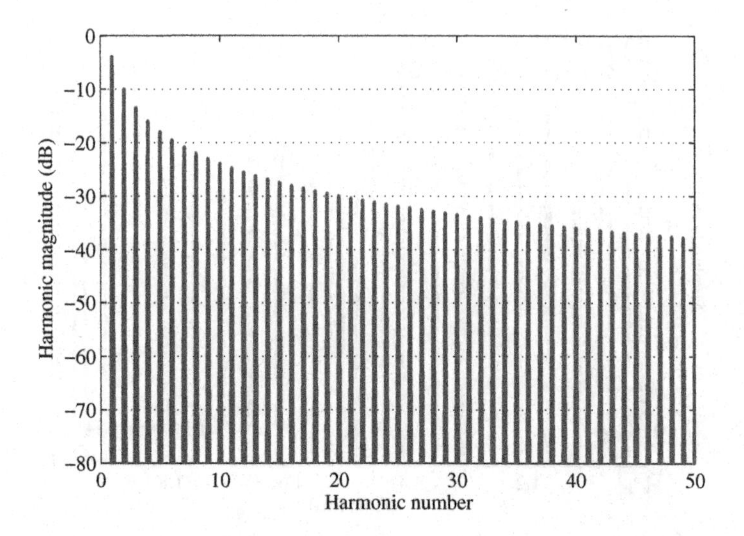

Fig. 2.16 Harmonic components of sawtooth carrier waveform.

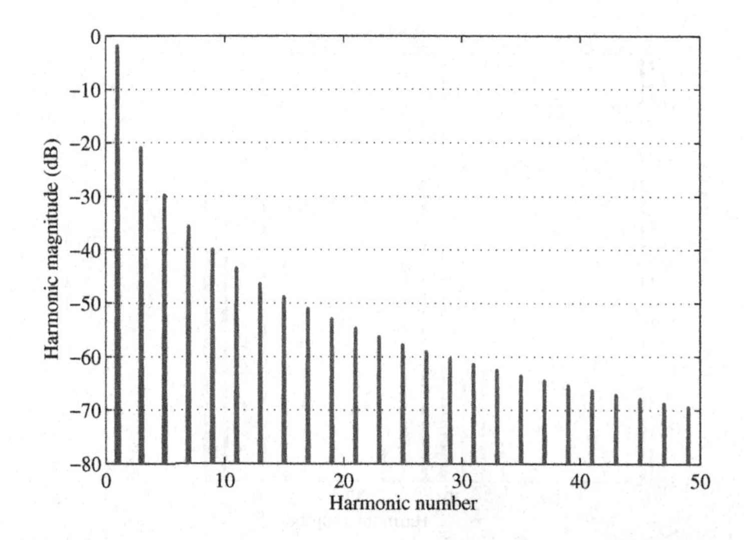

Fig. 2.17 Harmonic components of triangle carrier waveform.

model the carrier waveform employing the double Fourier integral analysis. This mathematical derivation is accurate but its complexity and procedure are extensive and tedious, and it has been only applied to a specific carrier waveform. Instead, the carrier waveform can be evaluated with

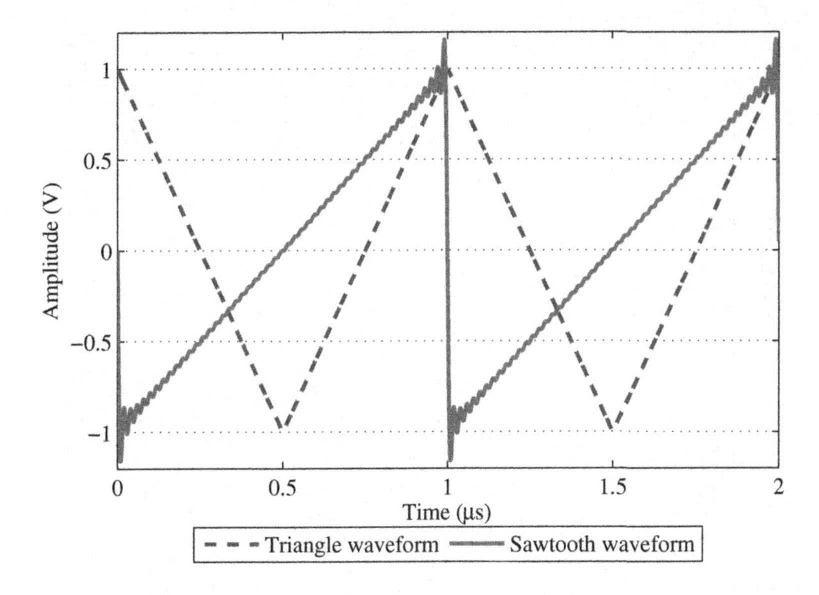

Fig. 2.18 Triangle and sawtooth waveforms with only fifty harmonic components.

the pulse-width modulation analysis by duty-cycle variation (ADCV) along with the Jacobi-Anger expansions [54].

The pulse-width modulation analysis by duty-cycle variation assumes that the input audio signal is constant within each carrier cycle, i.e. $\omega_c \gg \omega_o$, which is usually the case. The theory and derivations of the analysis by duty-cycle variation are detailed in Appendix A, and only the main results are shown in this chapter. Recall that a sawtooth/triangular wave carrier signal is constructed by an infinite sum of sinusoidal functions as expressed previously in Eq. (2.13) and Eq. (2.14), and since there are no unlimited bandwidth systems, the number of harmonics (i) in the carrier signals must be finite. Analyzing the latter case, a triangular wave carrier signal is plotted in Fig. 2.19(a) for different number of harmonic components. Observe that as the number of harmonic components increases, the triangular waveform approaches more to the ideal one. This phenomenon can be appreciated better in the magnified plot in Fig. 2.19(b). As a result of the finite number of harmonic components in the carrier waveform, the baseband harmonics of the pulse-width modulated signal, i.e. harmonic distortion, become non-zero and degrade the linearity performance of the class-D audio amplifier. Continuing with the analysis of the triangular waveform, there are two trivial cases regarding the number of harmonics

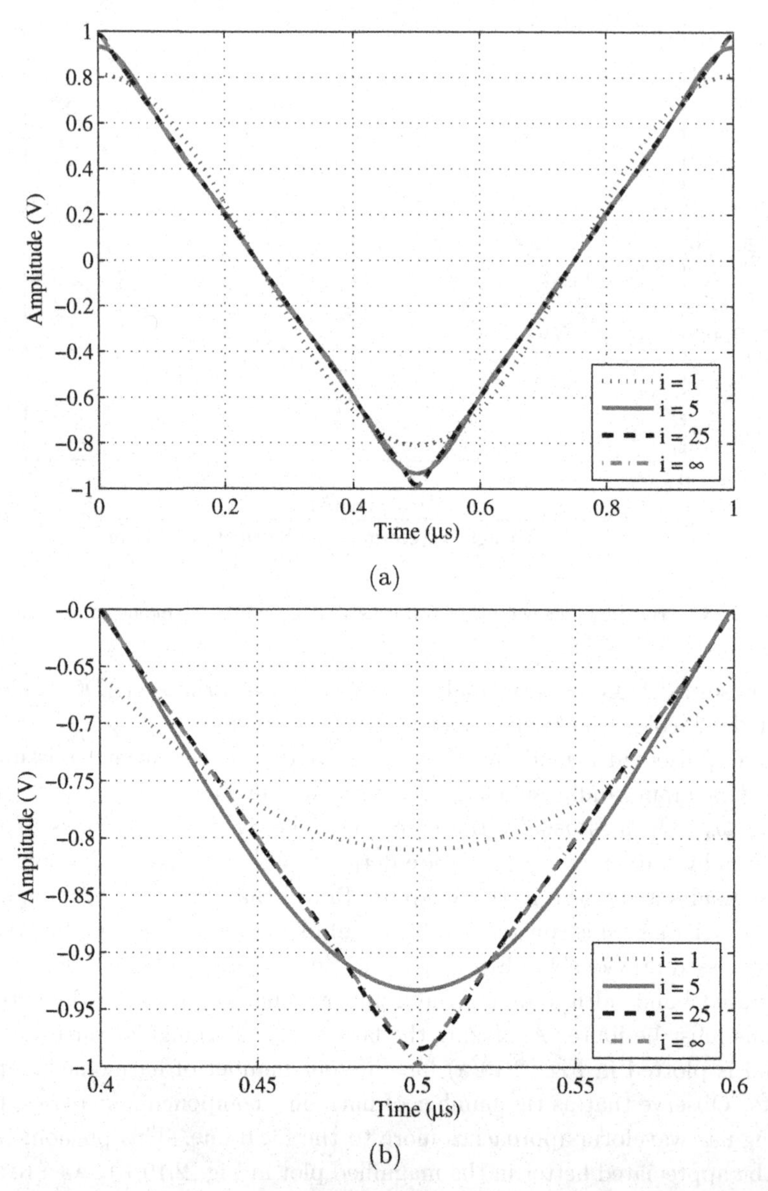

Fig. 2.19 (a) Triangular wave carrier signal for different number of harmonic components and (b) magnified view.

contained in the carrier signal, and the harmonic distortion they produce: (1) when the number of harmonics is infinite the triangle wave is ideal and the baseband harmonic components are zero, and (2) when the number of harmonics is one the triangle wave is a pure sinusoidal signal and the baseband harmonic components become dependents of the modulation index. Applying the analysis by duty cycle variation, as shown in Appendix A, to the second case, i.e. when the number of harmonics is one, the original pulse-width modulated signal expressed before in Eq. (2.11) becomes

$$v_{PWM}(t) = 2V_{DC} \arccos\left(\arcsin\left[2\sum_{n=1}^{\infty} \sin\left(n\frac{\pi}{2}\right) J_n\left(-\frac{1}{8}\pi^2 M\right)\cos\beta\right]\right)$$
$$+\frac{4}{\pi}V_{DC}\sum_{m=1}^{\infty} \sin\left(m \arccos\left[-\frac{1}{8}\pi^2 M\cos\left(\omega_o t + \theta_o\right)\right]\right)\cos\alpha \qquad (2.15)$$

where V_{DC} is the DC offset component, $J_n(\cdot)$ is the Bessel function of the first kind [54], and α and β are the arguments defined in Eq. (2.7) and Eq. (2.8).

Opposite to the pulse-width modulated signal with a pure triangle waveform derived previously in Eq. (2.11), the pulse-width modulated signal with a triangle waveform with only one harmonic component in Eq. (2.15) contains the fundamental tone with baseband harmonic components, whose magnitudes are proportional to the modulation index, which produce unwanted harmonic distortion.

Figure 2.20 illustrates graphically the baseband harmonic distortion, expressed as total harmonic distortion (THD), as a function of the modulation index M when the triangular carrier waveform contains only one harmonic component. This figure also shows the results of a simulated class-D amplifier with a single tone carrier signal in order to compare the results. Notice that the harmonic distortion increases as the modulation index increases as expected from Eq. (2.15).

The analysis by duty-cycle variation can be extended to any triangular carrier waveform with specific number of harmonic components (bandwidth), however, the only closed-form solutions exist when the number of harmonic components are $i = 1$ and $i = \infty$. The solution of the pulse-width modulated signal when the number of harmonic components in the triangular waveform carrier signal is $1 < i < \infty$ must be calculated numerically.

The numerical and simulated results of total harmonic distortion for triangular waveforms with different number of harmonic components are displayed in Fig. 2.21, where the analytical results are plotted with solid lines and the simulated results with markers. Observe that the distortion

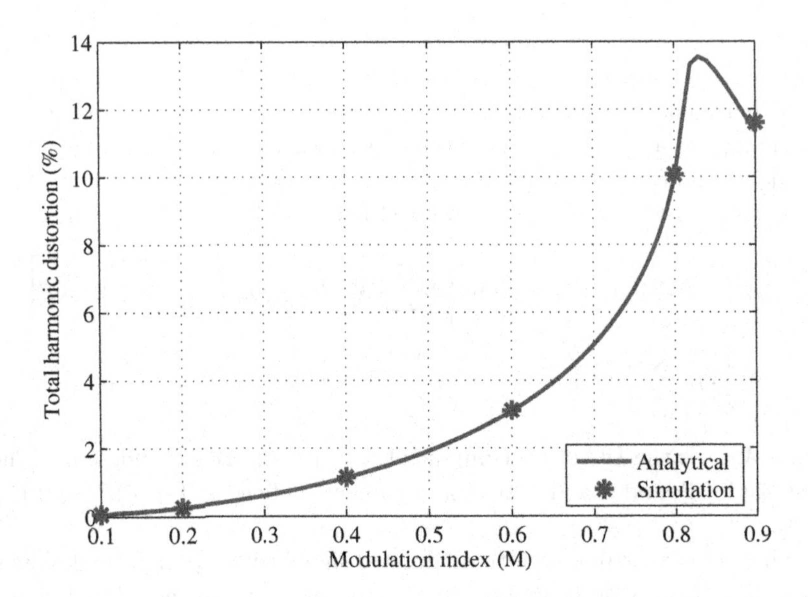

Fig. 2.20 Total harmonic distortion of class-D amplifier when triangle carrier waveform contains one harmonic component.

decreases as the number of harmonic components increases. Also, notice that the mathematical analysis predicts with very high accuracy the simulation results.

Therefore, the analysis of the pulse-width modulated signal by duty-cycle variation can be employed to limit the bandwidth of the class-D audio amplifier according to the desired harmonic distortion. In other words, the maximum total harmonic distortion allowed in a given class-D audio amplifier design will define the number of harmonic components of the triangular waveform, and consequently, the bandwidth of the system.

Harmonic distortion of class-D audio power amplifiers based on sawtooth modulation can also be analyzed using the duty-cycle variation method. In practice, a common carrier signal used to generate a naturally sampled pulse-width modulation is an exponential-shaped waveform due to its relatively simple implementation. The exponential-shaped waveform is usually generated by charging and discharging a simple RC integrator circuit with square pulses [3, 55, 56]. Figure 2.22 illustrates the pulse-width modulated signal based on an exponential-shaped carrier waveform.

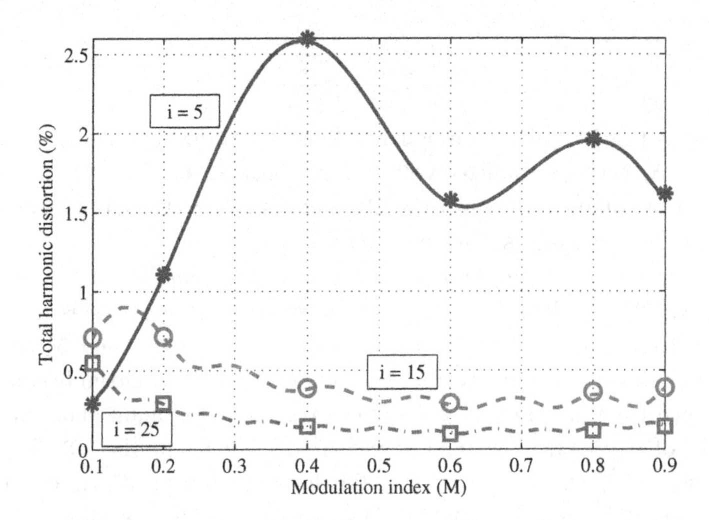

Fig. 2.21 Total harmonic distortion of class-D amplifier when triangle carrier waveform contains different number of harmonic components.

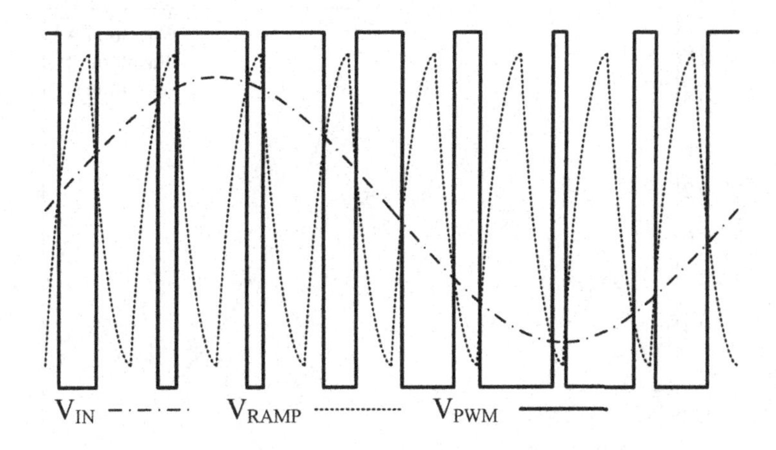

Fig. 2.22 Sine-exponential-shaped pulse-width modulation.

The exponential-shaped waveform f_e can be defined as

$$f_e(t) = \begin{cases} V_{DC}\left(2\dfrac{\exp\left(-\frac{t+\frac{T_c}{2}}{t_0}\right) - N_e}{(1 - N_e)} - 1\right) & \text{when } -\dfrac{T_c}{2} < t < 0 \\ V_{DC}\left(2\dfrac{1 - \exp\left(-\frac{t}{t_0}\right)}{(1 - N_e)} - 1\right) & \text{when } 0 \leq t < \dfrac{T_c}{2} \end{cases} \tag{2.16}$$

where

$$t_0 = -\frac{T_c}{2\ln N_e} \tag{2.17}$$

and V_{DC} is the DC offset component, T_c is the period of the carrier waveform, and N_e is the normalized voltage error between the maximum possible value of the square pulse into the RC network and the actual voltage at which the RC network is discharged (RC constant).

The carrier waveform defined by Eq. (2.16) resembles an exponential-shaped waveform when N_e is small because the voltage error is minimum and the charging time is maximum. On the other hand, when N_e is large, the carrier waveform approaches a triangular-shaped waveform because the voltage error is maximum and the linear region is dominant. For example, a family of exponential-shaped waveforms is shown in Fig. 2.23, where the values of N_e are 0.1, 0.3, and 0.9 for the waveforms A, B, and C, respectively. Observe the exponential shape of waveform A when $N_e = 0.1$, and the triangular shape of waveform C when $N_e = 0.9$.

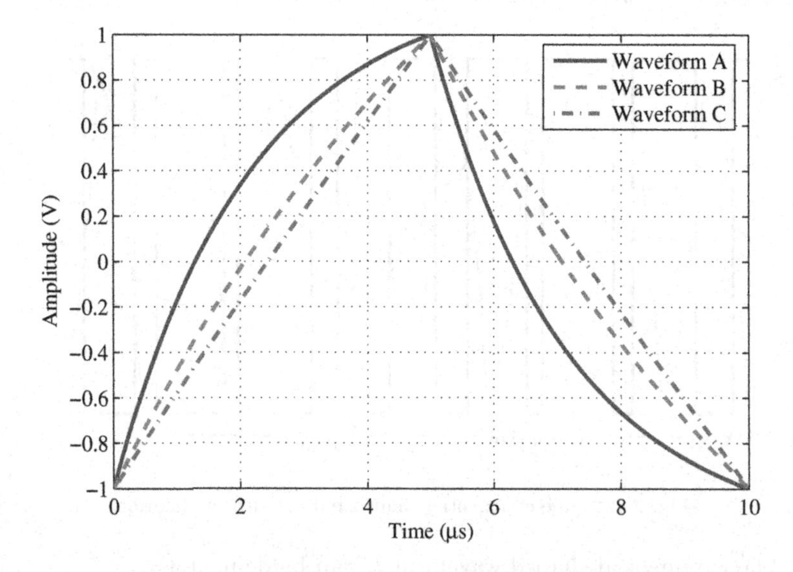

Fig. 2.23 Family of exponential-shaped waveforms for different values of N_e.

The pulse-width modulation analysis by duty-cycle variation can also be applied to the exponential-shaped carrier waveform, as detailed in Appendix A. In general, a pulse-width modulated signal with exponential-shaped carrier waveform can be expressed as the summation of its Fourier coefficients (a_m and b_m) as

$$v_{PWM}(t) = \frac{a_0}{2} + \sum_{m=1}^{\infty} (a_m \cos\alpha + b_m \sin\alpha) \tag{2.18}$$

where

$$a_0 = \frac{2}{\pi} V_{DC} \left[-t_0 \ln\left(1 - \frac{1}{2}[1 - N_e]\left[\frac{M}{V_{DC}}\cos\phi + 1\right]\right) + \pi + t_0 \ln\left(\frac{M}{2V_{DC}}\cos\phi[1 - N_e] + \left[\frac{(1 + N_e)}{2}\right]\right)\right], \tag{2.19}$$

$$a_m = \frac{2}{m\pi} V_{DC} \left[\sin\left[-mt_0 \ln\left(1 - \frac{1}{2}[1 - N_e]\left[\frac{M}{V_{DC}}\cos\phi + 1\right]\right)\right] - \sin\left[-m\pi - mt_0 \ln\left(\frac{M}{2V_{DC}}\cos\phi[1 - N_e] + \left[\frac{(1 + N_e)}{2}\right]\right)\right]\right], \tag{2.20}$$

$$b_m = \frac{2}{m\pi} V_{DC} \left[\cos\left[-mt_0 \ln\left(1 - \frac{1}{2}[1 - N_e]\left[\frac{M}{V_{DC}}\cos\phi + 1\right]\right)\right] - \cos\left[-m\pi - mt_0 \ln\left(\frac{M}{2V_{DC}}\cos\phi[1 - N_e] + \left[\frac{(1 + N_e)}{2}\right]\right)\right]\right], \tag{2.21}$$

$$\phi = \omega_o t + \theta_o \tag{2.22}$$

and V_{DC} is the DC offset component, ω_o is the audio signal angular frequency, θ_o is an arbitrary phase shift, and α is the argument defined in Eq. (2.7).

The harmonic distortion (baseband harmonic components) of the pulse-width modulated signal in Eq. (2.18) can be calculated for any given exponential-shaped carrier waveform. Figure 2.24 shows the total harmonic distortion of a class-D amplifier for the carrier waveforms shown in Fig. 2.23. Analytical results are plotted with solid lines and the simulation results, using the same carrier waveforms, are plotted with markers. Notice that the analytical results match the simulation results. Also, as expected, the distortion of the amplifier decreases as the carrier waveform approaches a triangular-shaped waveform.

A class-D audio power amplifier with three-level BTL topology is shown in Fig. 2.25. Notice that the three-level modulation is only possible in a differential architecture. Some commercial class-D audio power amplifiers [57] employ this modulation scheme. This modulation is also known as pulse-width modulation BD, or PWM BD [2]. It generates a ternary signal which produces no output pulses on absence of an audio signal at the input.

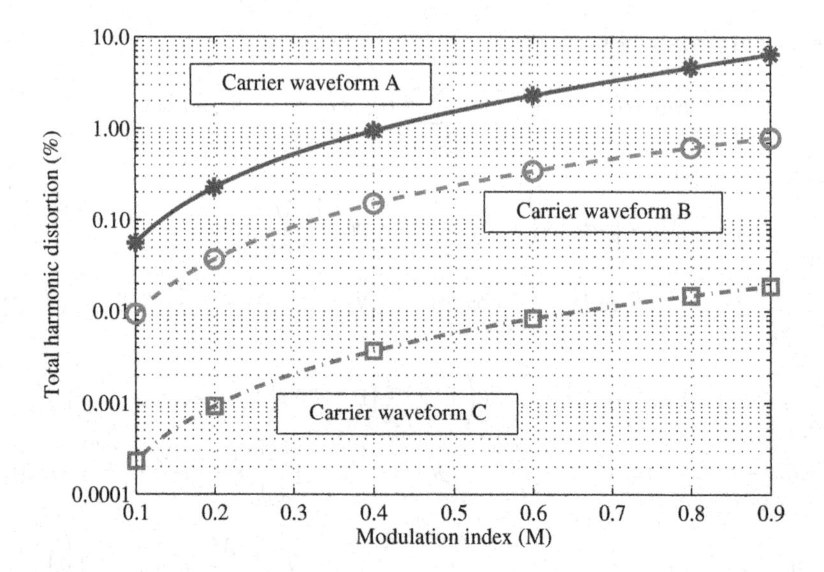

Fig. 2.24 Total harmonic distortion of class-D amplifier for different exponential-shaped carrier waveforms.

On the other hand, when the input audio signal is positive the pulse-width modulated signal is constructed with positive pulses, and during negative cycles of the input audio signal the modulation generates negative pulses. Also, the effective switching frequency, as it will be shown later, is twice of the reference carrier waveform. The higher the frequency, the better the filter attenuates unwanted frequencies and the residual ripple is minimized.

However, the three-level modulation scheme suffers of significant electromagnetic interference (EMI) from the common mode output signal because this signal swings rail-to-rail at the amplifier switching frequency. Then, the class-D amplifier requires to be mounted as close as possible to the loudspeaker. Another potential problem of the three-level modulated class-D amplifier is the crossover distortion. This problem can be eliminated by designing the power stage logic so that the class-D amplifier puts out very narrow alternating positive and negative pulses in the absence of an input signal [2, 3, 58].

Figure 2.26 details the generation of the three-level pulse-width modulated signal. The audio signal and the carrier wave generate two complementary out-of-phase digital signals (PWM+ and PWM−) whose differential voltage creates a third modulation level.

The three-level modulated signal is created because a fully-differential

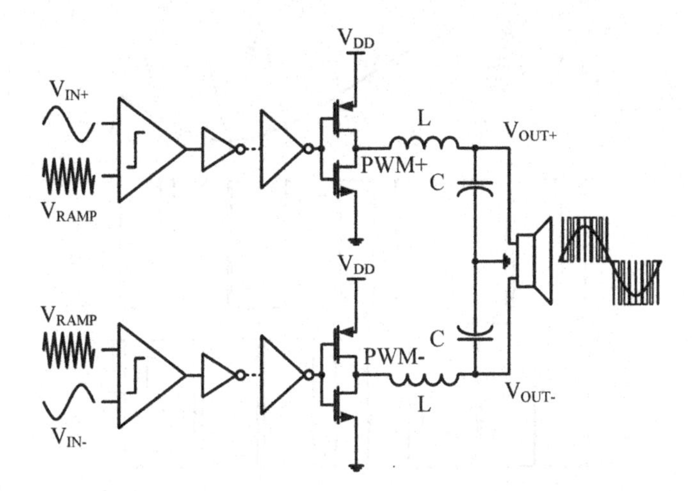

Fig. 2.25 Three-level bridge-tied-load class-D audio power amplifier.

audio input signal is compared with the same triangular waveform carrier. In other words, one of the pulse-width modulated outputs is inverted in time within one sampled period.

Three-level pulse-width modulated signals can also be generated using the other carrier waveforms (sawtooth waveform, exponential-shaped waveform). Moreover, the harmonic components of all the modulation schemes can be analyzed by using the double Fourier integral analysis and/or the analysis by duty-cycle variation along with the Jacobi-Anger expansions [54]. For example, the ternary pulse-width modulated signal with a sawtooth waveform carrier (trailing-edge naturally sampled modulation) is given by

$$\begin{aligned} v_{PWM}(t) &= 2V_{DC}M\cos\phi \\ &\quad + \frac{4}{\pi}V_{DC}\sum_{m=1}^{\infty}\sum_{\substack{n=-\infty \\ (n\neq 0)}}^{\infty}\frac{1}{m}J_n(m\pi M)\sin\left(n\frac{\pi}{2}\right)\cos(\alpha\ +\ \beta) \end{aligned} \tag{2.23}$$

and the ternary pulse-width modulated signal with a triangular waveform carrier (double-edge naturally sampled modulation) is given by

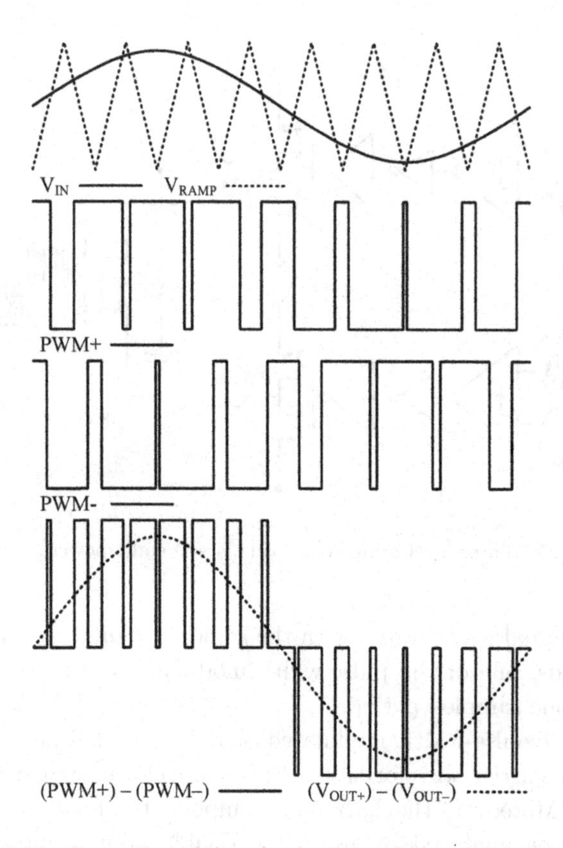

Fig. 2.26 Generation of three-level pulse-width modulated signal.

$$
\begin{aligned}
v_{PWM}(t) &= 2V_{DC}M\cos\phi \\
&\quad + \frac{8}{\pi}V_{DC}\sum_{m=1}^{\infty}\sum_{n=-\infty}^{\infty}\frac{1}{2m}J_{2n-1}(m\pi M) \\
&\quad \times \cos([m+n-1]\pi)\cos(2\alpha+[2n-1]\phi) \qquad (2.24)
\end{aligned}
$$

where V_{DC} is the DC offset component, M is the modulation index, $J_{(\cdot)}(\cdot)$ is the Bessel function of the first kind [54], and α, β, and ϕ are the arguments defined in Eq. (2.7), Eq. (2.8), and Eq. (2.22). Figure 2.27 shows the time domain representation of the three-level pulse-width modulated signal in Eq. (2.23) with coefficients $m = n = 50$, modulation index $M = 0.9$, $V_{DC} = 1\ V$ and $\omega_c / \omega_o = 21$, and Fig. 2.28 illustrates its harmonic components. Observe that all harmonics where coefficient n is even are canceled and the harmonic composition is significantly less than the two-level pulse-width modulated signal plotted in Fig. 2.12.

The three-level pulse-width modulated signal in Eq. (2.24) with coefficients $m = n = 50$, modulation index $M = 0.9$, $V_{DC} = 1\ V$ and $\omega_c/\omega_o = 21$ is shown in Fig. 2.29, and Fig. 2.30 plots its harmonic spectra. Note that the indices m and n produce only even carrier multiples ($2m$) with odd sideband harmonics ($2n - 1$), and the effective switching frequency of the carrier waveform is doubled. As in the previous case, the number of harmonic components have significantly decreased when compared to the two-level pulse-width modulated signal shown in Fig. 2.15.

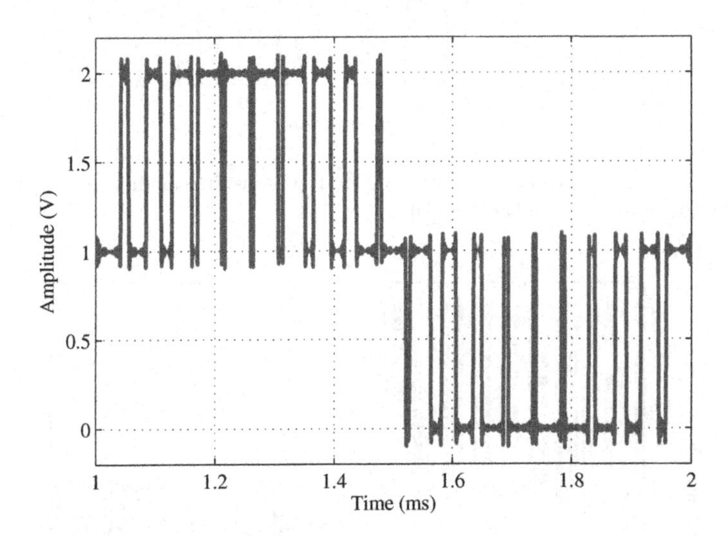

Fig. 2.27 Three-level pulse-width modulated signal with trailing-edge naturally sampled modulation.

The analysis of two-level pulse-width modulation harmonic distortion can also be extended to three-level pulse-width modulation signals, and even multi-level pulse-width modulation schemes, to determine the necessary characteristics, i.e. bandwidth, number of harmonic components, etc., of the carrier waveform for a given linearity specification.

Class-D audio power amplifiers are inherently open-loop systems, and, as it has been seen before, their linearity performance relies heavily on the quality of the carrier waveform. Moreover, they provide very poor power-supply rejection ratio (PSRR) and do not have gain control. The open-loop architecture offers simple, efficient, and smaller area [55]. In practice, they

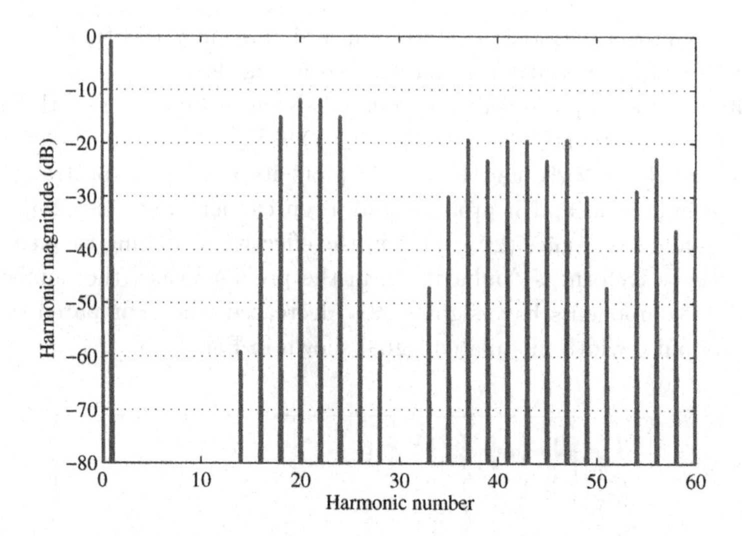

Fig. 2.28 Harmonic components of three-level pulse-width modulated signal with trailing-edge naturally sampled modulation.

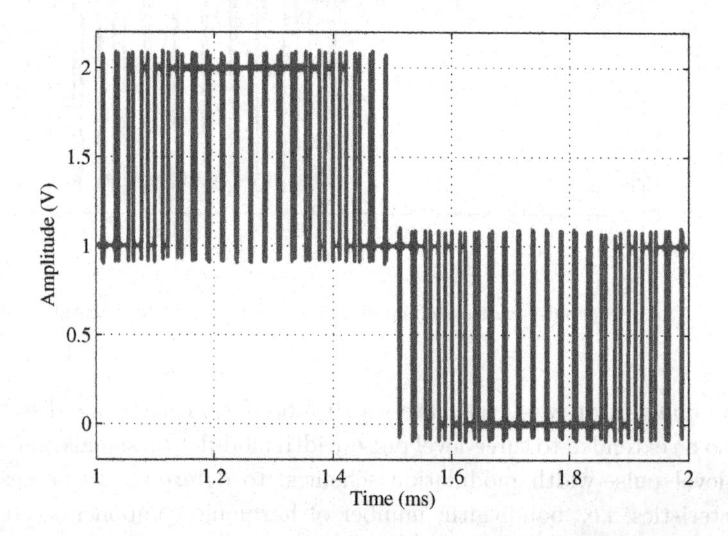

Fig. 2.29 Three-level pulse-width modulated signal with double-edge naturally sampled modulation.

are used for low quality audio applications [2].

Negative feedback is often used around the conventional class-D amplifier to improve its performance [3]. The closed-loop architecture provides

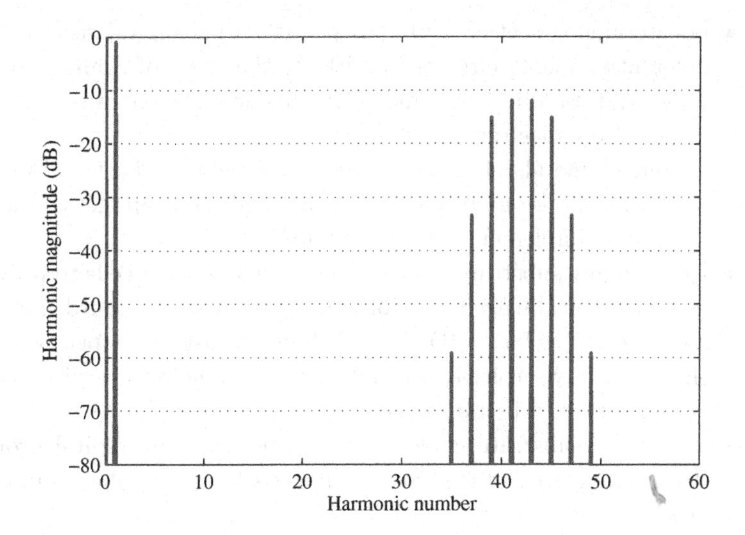

Fig. 2.30 Harmonic components of three-level pulse-width modulated signal with double-edge naturally sampled modulation.

an improved topology with more robustness to non-ideal effects. The negative feedback increases linearity and improves the power-supply rejection ratio [3, 56, 59, 60]. The closed-loop class-D architectures will be detailed in Chapter 3.

2.5 Layout and printed circuit board recommendations

The layout phase of class D audio power amplifier is crucial in order to get a good performance during the testing stage. Analog section should be laid off using standard layout techniques such as common-centroid arrangements and use of dummy components for best matching [61].

In addition, special effort must be taken when the output power stage is being designed because the coupling of substrate noise can destroy the analog circuitry performance, and narrow tracks can attenuate the efficiency of the amplifier. Additional layout techniques as guard-rings and metal/via resistance minimization should be employed to reduce the effect of substrate noise in the amplifier.

A suggested list of guidelines for laying out class-D audio power amplifiers is shown below.

- Locate analog section as far as possible from digital section, use differential analog circuits to mitigate the effect of common-mode noise, and use dummy elements for analog section (transistors, resistors and capacitors).
- Layout of metals should be transversal between layers. Bottom metals should be used to connect local cells and top metals should be used to implement the power grids.
- Use as many substrate contacts as possible in local cells to provide a homogenous bulk voltage for transistors. Use P+ guard ring and N+ guard ring for NMOS and PMOS transistors, respectively. If the cell is very sensitive, use of dual rings can help to provide better isolation from substrate noise.
- Even that substrate could be the same, separate digital ground from analog ground and routed them as far as possible from each other.
- Use as many contacts as possible in power grids and use wide tracks for power connections to minimize sheet and via resistances.
- Put guard rings as close as possible to circuits, it will reduce the resistance between a noisy circuitry and a ground path.
- Use as many as possible number of pads for digital section, in such way, the bonding inductance will be minimized, and use different frame with ESD circuit protection for analog and digital section to separate noisy common connections.
- If possible, use dedicated guard rings around analog and digital sections, each one of them connected with a dedicated path to the power supply. Such guard rings must be as wide as possible.

Once the integrated circuit has been fabricated, the printed circuit board (PCB) should follow a good design to minimize the risk of degrading the performance of class D audio power amplifiers. There are three main areas concerning the PCB design: (1) the ground plane, (2) the power plane, and (3) the inputs and the outputs.

A solid ground plane works as well as other types of grounding schemes because the system operates at relatively low frequency. A solid ground plane also helps to assist in the dissipation of heat, keeping the class D audio amplifier relatively cool and negating the need for an external heat sink. Additionally, the ground plane acts as a shield to isolate the power pins from the output and to provide a low-impedance ground return path. It is important that any components connecting an IC pin to the ground plane be connected to the nearest ground for that particular pin.

The power plane contains two main different sections, the analog power pins and the output stage power pins. In general, the power traces must be kept short and the decoupling capacitors should be placed as close to the power pins as possible. The analog plane supplies power for sensitive circuitry and is the most sensitive pin of the device. Therefore, it must be kept as noise free as possible. The output stage power plane is not as sensitive to noise as the analog power plane but its design must be done carefully to minimize ground loops and to provide very short ground return paths.

Figure 2.31 shows a bad design for the power plane in the output power path of a class D audio power amplifier because the loop area, Area $=$ A $\times$ B, is large. On the other hand, Fig. 2.32 shows a very good design of the output power loop which minimizes the loop area to Area $=$ (A $\times$ B) $-$ (a $\times$ b). Finally, the input and output power planes must be separated. The loudspeaker traces should be kept as short as possible to reduce noise pickup. The bias network have almost no current flowing through them and then there are no special consideration for the layout of those traces. Standard layout practices will apply. The trace lengths between the output pins and the LC filter components must be minimized. The traces to the inductors should be kept short and separated from the input circuitry as much as possible. All high-current output traces should be wide enough to allow the maximum current to flow freely. Failure to do so creates excessive voltage drops, decreases the efficiency, and increments the distortion.

2.6 Typical applications

The high power efficiency of class-D amplifiers makes them suitable for a wide range of applications such as boom boxes or portable stereo systems, portable video players, hearing aids, notebook computers, tablet computers, smart phones, etc. The characteristic low power dissipation in the CDA extends their applications to audio systems that are not battery powered but where reduced weight and size are important. These applications are speaker systems for amusement parks, stadiums, home theater, televisions, car audio, etc. The CDA applications can be broadly classified in two categories: 1) low power applications targeted for battery-powered portable devices, and 2) medium to high power applications targeted for reduced heat dissipation audio systems. The low power applications require output powers less than 3 W, and their main focus is high power efficiency

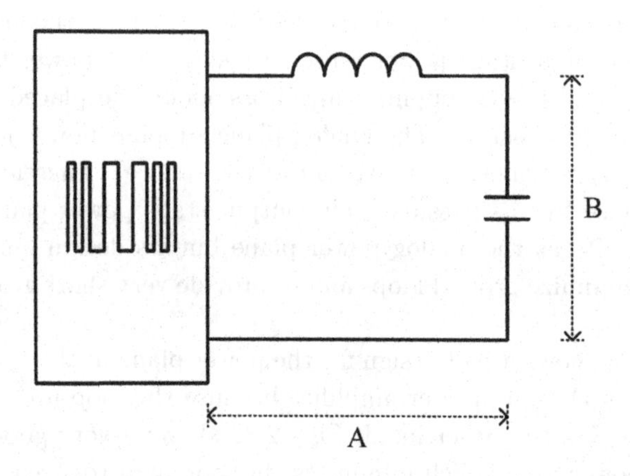

Fig. 2.31 Diagram of amplifier and load with large loop area.

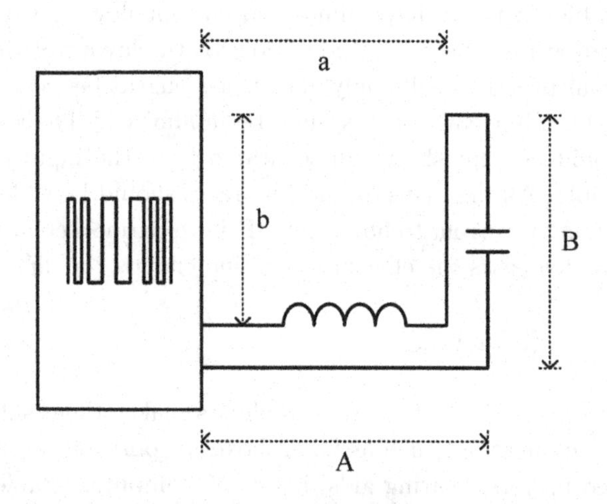

Fig. 2.32 Diagram of amplifier and load with small loop area.

($\eta > 90\%$) and high linearity (THD + N $< 0.1\%$). The medium to high power applications require output powers from 10 W up to 3600 W with small heat sinks and compact size, and their main focus is high efficiency ($\eta > 80\%$) at peak output power for reduced weight and size.

For the purpose of this book, the CDA for low-power applications will be targeted since they are more compatible to the semiconductor voltage limits in integrated circuits, instead of discrete components in the case of medium to high power applications. However, the principles and theory that will be discussed can be applied to all class-D amplifiers in general, and can be extrapolated to high-power applications if needed.

2.6.1 *Commercial class-D audio amplifiers typical specifications*

To understand the trends in class-D audio amplifiers for mobile devices, state-of-the-art commercial CDA specifications are shown in Table 2.1. It can be observed that the specification for supply voltage corresponds to the battery maximum supply voltage; for the lithium-ion batteries used in mobile devices, the standard voltage ranges from 4.8 V when fully charged to 2.7 V when almost discharged.

Table 2.1 Typical specifications for commercial class-D audio amplifiers.

Parameter	[57]	[62]	[63]	[64]	[65]
Supply (V_{DD})	5	5	5	5	3.6
I_Q (mA)	7	1.42	6.5	4.2	2.7
P_Q (mW)	35	7.1	32.5	21	9.7
Efficiency(%)	85	90	85	90	85
THD+N (%)	0.65	0.02	0.02	0.08	0.01
$P_{o,max}$ (W)	1.2	1.7	1.4	1.7	2.3
PSRR (dB)	65	88	85	93	88
SNR (dB)	83	98	96	89	97
F_{SW} (kHz)	250	192	420	300	300

Another observation in terms of audio performance is that a THD+N smaller than 0.1 %, SNR higher than 80 dB, PSRR higher than 60 dB, and efficiency higher than 80 % are required to be competitive. The power dissipation is proportional to the maximum output power provided ($P_{o,max}$), but the smaller the quiescent power (P_Q), the longer the battery will last. Typical switching frequencies (F_{SW}) are in the range of 200 kHz to 400 kHz.

Chapter 3

Closed Loop Architectures for Class-D Amplifiers

3.1 Closed loop architectures

Open loop class-D amplifier (CDA) architectures are cost effective and simple to implement, as shown in Figs. 2.1 and 2.2. However, the absence of error correction makes them too sensitive to variations in components, timing errors, and supply noise. The main applications that leverage the low power and simplicity of the open loop CDA are toys, smoke alarms, and buzzers. To achieve outstanding audio performance in a CDA, closed-loop architectures are typically used where the negative feedback mechanism helps to correct errors in the amplification process. The close loop transfer function is typically expressed as,

$$\frac{V_O(s)}{V_I(s)} = \frac{A_{ol}(s)}{1 + A_{ol}(s) \cdot \beta(s)} \cong \frac{A_{ol}(s)}{LG(s)} \cong \frac{1}{\beta(s)} \tag{3.1}$$

where the $A_{ol}(s)$ represents the open loop gain of the system, and β is the feedback factor. It is assumed that the close loop gain is large (e.g. $LG(s) = A_{ol}(s) \cdot \beta(s) \gg 1$). Two important aspects can be noticed from the feedback mechanism: 1) if the β factor is chosen as a linear gain, then the close loop system will have a linear behavior; 2) any non-linearity in $A_{ol}(s)$ will be attenuated by $LG(s)$.

The general structure for a close loop CDA is shown in Fig. 3.1 where the darker blocks comprise the closed-loop gain of the system. $A_{ol}(s)$ is given by the small signal models of the compensator, modulator, and output stage. It can be noticed that the output filter and speaker are outside of the feedback loop. Thus, their errors and non-linearities would not be corrected by the feedback. Also, the feedback signal is the switching signal of the class-D output stage, meaning that it is a high frequency square wave at full swing with a high number of harmonics. Thus, the compensator, modulator, and class-D output stage have to process all the frequency harmonics of the feedback signal, increasing the design complexity of each block.

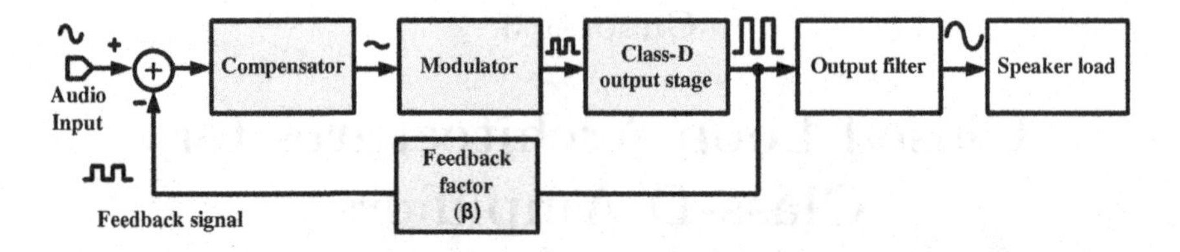

Fig. 3.1 General closed-loop CDA architecture.

The closed-loop CDA architecture operates as follows: the compensator function is to extract and filter the error signal coming from the difference between the audio input signal and the feedback signal. Also, this block has to provide gain in the loop to attenuate distortion and errors at the output, and ensure stability in the system. The compensator is typically implemented using an integrator chain where the order of the compensator is proportional to the number of integrators in the chain. The modulator function is to process the output of the compensator and implement the desired modulation scheme. This modulation is typically implemented with a high frequency carrier. The modulator output signal is then passed through a chain of digital inverters that increase their output drive with each stage. Finally, the output of the inverters have to charge and discharge the large gate capacitors of the class-D output switches to be able to turn them on or off. The output switching signal is then applied to the feedback factor (β) and returned to the input of the compensator to close the loop.

Another alternative for the feedback signal is shown in Fig. 3.2. $A_{ol}(s)$ is given by the small signal models of the compensator, modulator, output stage, and output filter. This provides the advantage of including the output filter inside the feedback loop, correcting the non-linearities in the filter components. Also, the output filter removes the high frequency components of the feedback signal, leaving only the low frequency information of interest. This relaxes the design complexity of the compensator, modulator, and class-D output stage. The main drawback is that the output filter typically has two poles, requiring a complex compensation scheme to make the system stable. Compensators using non-linear control provide a possible solution to ensure stability and fast transient response when the output filter is included in the loop, as will be shown in Chapter 6.

It is worth noticing that the closed-loop architecture in Fig. 3.2 is also

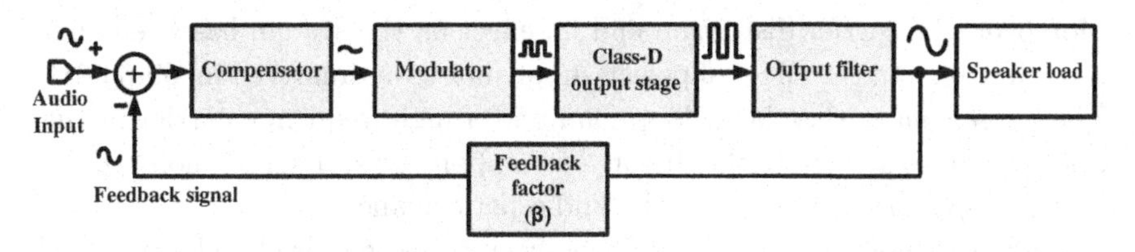

Fig. 3.2 Closed-loop CDA architecture with alternative feedback, including output filter.

used in DC-DC Buck converters [43, 66]. However, the goal is to regulate the output voltage under large load transients to reduce the output voltage ripple, and using the compensator to stabilize the system with a fast transient response. Since the output signal is a constant voltage, the bandwidth of the loop is designed to react to the fastest load change, which in modern microprocessors could be in the range of tens of nanoseconds.

In some CDA for mobile devices where boosted supplies are needed to deliver more output power, the main purpose of β is to attenuate the high voltage swing of the output to make it compatible with voltage levels tolerable by the compensator. A common choice for this feedback factor is $\beta = 1$ since the compensator is also connected to the battery supply as well as the output stage. Another consideration for the feedback factor is that its errors or non-linearities will appear directly at the output, as expressed in Eq. (3.1). Thus, β is typically implemented as a linear resistive voltage divider.

One of the most important design choices for the closed-loop CDA architecture is the switching frequency F_{SW}, since its value affects directly the power dissipation of the system, as expressed in Eq. (2.2), and the linearity of the system. The Nyquist theorem [67] establishes that the minimum sampling frequency needed to recover accurately a sampled input signal has to be at least 2 times the frequency bandwidth of the desired signal. For audio, the frequency bandwidth is 20 kHz. Therefore, the minimum sampling frequency to satisfy the Nyquist theorem is ideally 40 kHz. However, this condition does not take into account the non-idealities, such as finite slew-rate, of the implemented carrier signal or modulator. To avoid any intermodulation distortion caused by the aliasing of the carrier frequency and the audio signal high frequency components, a typical rule-of-thumb is to choose F_{SW} at least 10 times larger than the desired bandwidth.

The general closed-loop CDA has been analyzed for intermodulation distortion (IMD) in time domain [68], and in frequency domain [69].

Moreover, the carrier distortion and its effect on the system has been analyzed in [70], and the effect of power-supply noise was analyzed in [59,60,71]. The agreement is that large loop gain and a high-frequency carrier in the system help to attenuate the distortion components and supply noise of the closed-loop system, improving the audio performance.

Different modulation schemes have been proposed for closed-loop CDA architectures to achieve high efficiency and good audio performance using modulation techniques such as PWM [44, 47, 48, 56, 72–76], sigma-delta modulation (SDM) [45, 77], or self-oscillating modulation (SOM) [78–83]. Each modulation scheme has its advantages and disadvantages, depending on the implementation and application. A brief review for each of the main modulation techniques used in closed-loop CDA architectures is presented next.

3.2 Pulse-width modulation

The closed-loop pulse-width modulation (PWM) CDA architecture operates in a similar way as the open-loop case shown in Fig. 3.3, where a high frequency carrier signal V_C is compared to a low frequency signal V_I to generate the modulated squarewave signal amplified by the class-D output stage V_{SW}. The difference is that a compensator block is added, as shown in Fig. 3.4, to correct the error (V_e) between the input and the feedback signal, and to provide gain and stability to the closed-loop system.

The main design parameter for a first order compensator is the time constant τ_I, which will determine the unity gain frequency UGF $= 1/(2\pi \cdot \tau_I)$, also known as gain-bandwidth product (GBW), of the system. Its value selection depends on several tradeoffs between the PWM carrier frequency, CDA implementation silicon area, amplifier power, linearity, and noise. To understand the tradeoffs involved in the design of the closed-loop PWM CDA architecture, first, second, and third order compensator design examples will be discussed.

The first-order compensator is typically implemented using an active integrator with an operational amplifier. The advantage of the first order compensator is that it only has one pole, ensuring a stable system if UGF $\ll F_{SW}$. The small signal transfer function for the first-order ideal compensator $G_{C,1st}(s)$, modeling the comparator and output stage as an unity linear gain, is expressed as,

$$G_{C,1st}(s) = \frac{V_{SW}(s)}{V_e(s)} = \frac{1}{s \cdot \tau_I} \tag{3.2}$$

where τ_I is the integration time constant, and assuming $V_{I,max} = V_{SW,max} = V_{C,max} = V_{DD}$. In reality, the comparator and output stage, linearized as a gain block, will have a gain different than one since the compensator, carrier signal generator, and comparator are operated from a lower voltage supply than the output stage to reduce power consumption. The comparator and output stage linearized gain and modeling will be detailed in Chapter 5.

For very small frequencies (e.g. $s = 0$), the magnitude of $G_{C,1st}(s)$ is ideally infinite, while for very high frequencies (e.g. $s = \infty$), the magnitude is zero, and for frequencies between these two points, the magnitude has

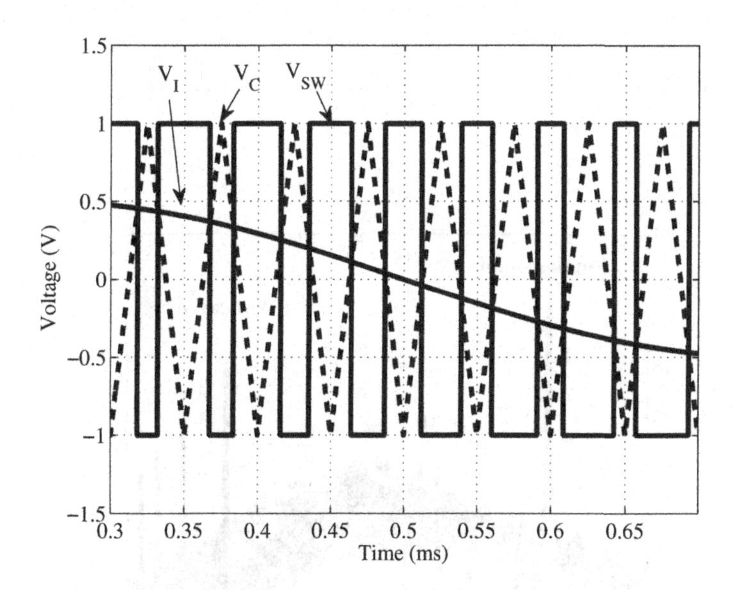

Fig. 3.3 Open-loop PWM waveforms.

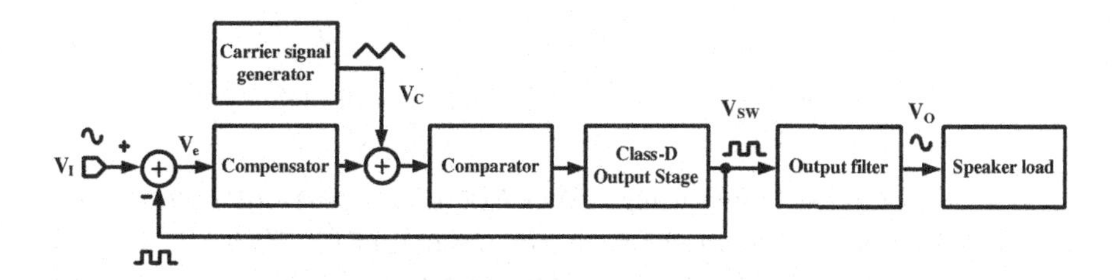

Fig. 3.4 Closed-loop PWM CDA architecture.

a roll off or slope of −20 dB for each decade in frequency. However, the implementation of the active integrator will be limited by the finite low frequency gain and finite GBW of the operational amplifier.

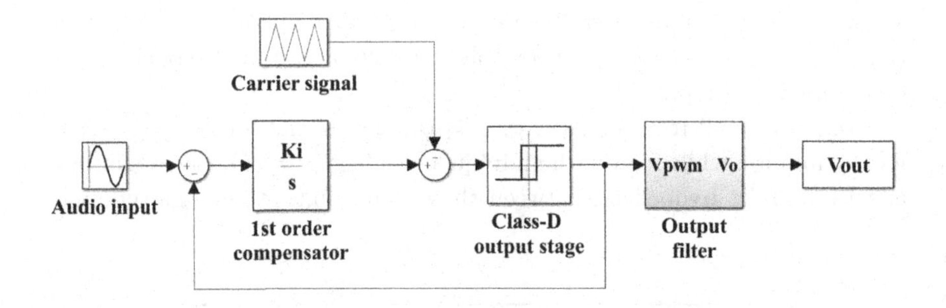

Fig. 3.5 Simulink model for 1st order PWM CDA architecture.

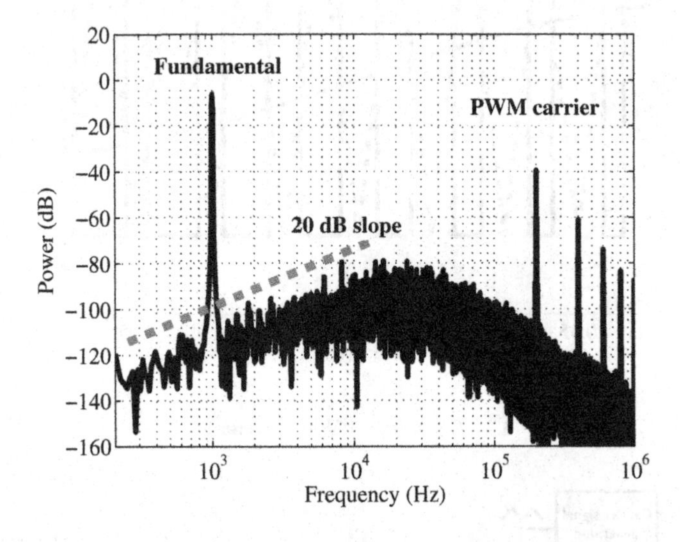

Fig. 3.6 Output spectrum for 1st order PWM CDA with UGF=50 kHz.

A system level example using MATLAB© for a first-order close loop PWM CDA can be simulated for an input signal of 0.5 V_{RMS} at 1 kHz, F_{SW} = 200 kHz, a Butterworth second order low pass filter with cut-off frequency of 20 kHz, and UGF = 50 kHz. The Simulink model used for simulation is illustrated in Fig. 3.5 where $Ki = 2 \cdot \pi \cdot UGF$. The frequency spectrum of the output signal is shown in Fig. 3.6, where the input signal fundamental frequency and the carrier signal are clearly visible. Also, the roll-off of the first-order compensator is noticeable in the audio frequency bandwidth. The third harmonic of the input signal is -90 dB from the fundamental harmonic. From the compensator open-loop frequency response, we can observe that this UGF choice provides a low frequency gain of 66 dB with a roll-off of -20 dB/dec, showing a single-pole system behavior.

If the UGF is chosen at lower frequencies, then the gain in the loop will be reduced; consequently, the distortion will increase. This effect can be observed in Fig. 3.7 where a UGF = 25 kHz was chosen while keeping the other parameters the same. It can be observed that the third harmonic is -84 dB from the fundamental harmonic of the signal; this -6 dB degradation is expected since the UGF was reduced in half and the gain in the loop also reduced in half. This choice of UGF provides a low frequency gain of 60 dB in the loop. Also, the high order harmonics are not attenuated due to the low UGF of the compensator and the overall THD+N is degraded.

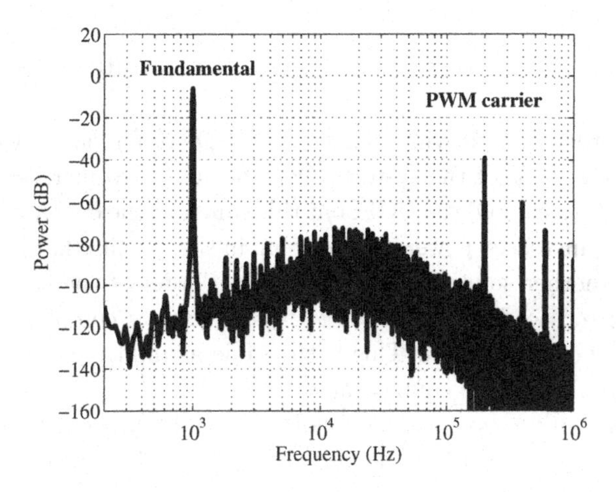

Fig. 3.7 Output spectrum for 1st order PWM CDA with UGF = 25 kHz.

If the UGF is chosen at higher frequencies, then the UGF of the system will start to get close the switching frequency and the stability of the system will degrade, increasing the IMD of the system and degrading the THD+N. To verify this, the UGF was increased to 100 kHz, that is the theoretical maximum UGF according to the Nyquist criteria for this example, and the output spectrum is shown in Fig. 3.8. This choice of UGF provides a higher low frequency gain of 72 dB in the loop, attenuating an extra 6 dB in comparison to the original case, but the IMD components at the PWM carrier frequency degrade the overall linearity.

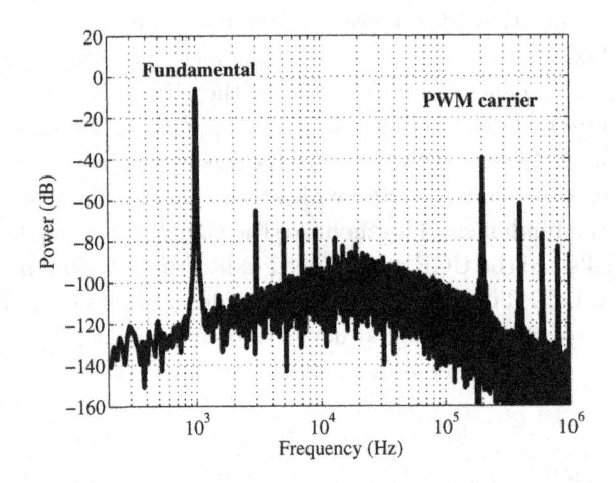

Fig. 3.8 Output spectrum for 1st order PWM CDA with UGF = 100 kHz.

The first order system will enhance its THD+N for larger UGF, but restricted by the F_{SW} of the system. One solution is to increase F_{SW} to have more room to expand the UGF, but this comes at the expense of design complexity and increased power consumption in the circuits that implement the system. Another solution is to increase the order of the compensator by adding more cascaded integrators at the expense of complex design to ensure stability of the system. This is the same as implementing a $n-th$ order low-pass filter as the compensator.

The second order compensator follows the same procedure to choose the integrators time constant. The only difference is that the compensator now has two integrators in cascade, providing -40 dB slope in the desired bandwidth. However, the stability of the system starts to degrade since the system has two poles in the origin with a phase shift of 180°. To ensure stability, an extra zero has to be added to the compensator and its placement in frequency will affect the dynamics of the second order system. The small signal transfer function for the second-order ideal compensator $G_{C,2nd}(s)$, modeling the comparator and output stage as a unity linear gain, is expressed as,

$$G_{C,2nd}(s) = \frac{V_{SW}(s)}{V_e(s)} = \frac{1 + s \cdot \tau_Z}{(s \cdot \tau_I)^2} = \frac{1 + s \cdot K_Z \cdot \tau_I}{(s \cdot \tau_I)^2} \tag{3.3}$$

where τ_Z represents the zero time constant, and the constant $K_Z = \tau_Z/\tau_I > 1$ is typically used to determine the zero frequency location as a function of the integrator pole location to ensure stability in the system.

The UGF is now affected by the zero frequency location, especially by the choice of K_Z, and it can be estimated by finding the frequency at which the magnitude of Eq. (3.3) is unity as,

$$UGF_{PWM,2nd} \cong \frac{K_Z}{2\pi \cdot \tau_I} = \frac{\tau_Z}{2\pi \cdot \tau_I^2}. \tag{3.4}$$

To ensure a stable close loop system, the phase margin (PM) of $G_{C,2nd}(s)$ needs to be larger than 40° to avoid ringing in the output signal. The main purpose of the extra zero is to introduce a phase boost to satisfy the PM requirement. The choice of K_Z will affect the PM of the system as,

$$PM_{PWM,2nd} = \tan^{-1}(K_Z^2). \tag{3.5}$$

Using the extra zero for compensation, we can extend the UGF beyond the value of the first order compensator. A large K_Z would increase the PM and UGF of the system. However, the UGF cannot get too close to the F_{SW} to avoid IMD and distortion. These effects are shown in the Bode plot of Eq. (3.3) for three choices of $K_Z = 0, 3, 10$, as illustrated in Fig. 3.9.

Small values of K_Z do not provide enough PM in the system, but for large values of K_Z, the system UGF is too large and close to the switching frequency that the closed-loop system would be unstable.

The Nyquist plot provides insight on the stability of the close loop system, as observed in Fig. 3.10 for a F_{SW}=200 kHz. A careful choice of K_Z is needed, taking into account the value of τ_I and F_{SW}. A rule-of-thumb choice for K_Z is between 1.2 and 3, introducing enough phase boost to keep the UGF constrained.

The benefit of increasing the loop gain by increasing the order of the compensator, as expressed in Eq. (3.1), is a better audio performance. To demonstrate this, the same system level parameters of the first-order compensator will be used for the simulation of a second-order closed-loop PWM CDA with a UGF = 50 kHz and $K_Z = 2$. The Simulink model used for simulation is illustrated in Fig. 3.11. The resulting output spectrum is shown in Fig. 3.12, where the third harmonic of the input signal is -96 dB below the fundamental, and the in-band noise floor has a 40 dB/decade attenuation.

The distortion and noise floor have been attenuated even more,

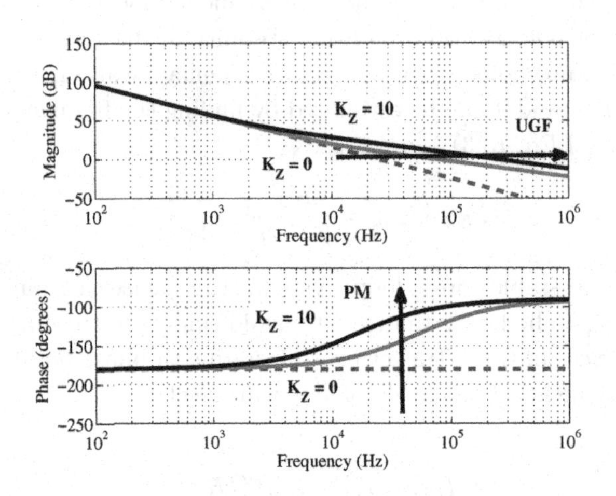

Fig. 3.9 Bode plot for second order PWM CDA for different K_z values.

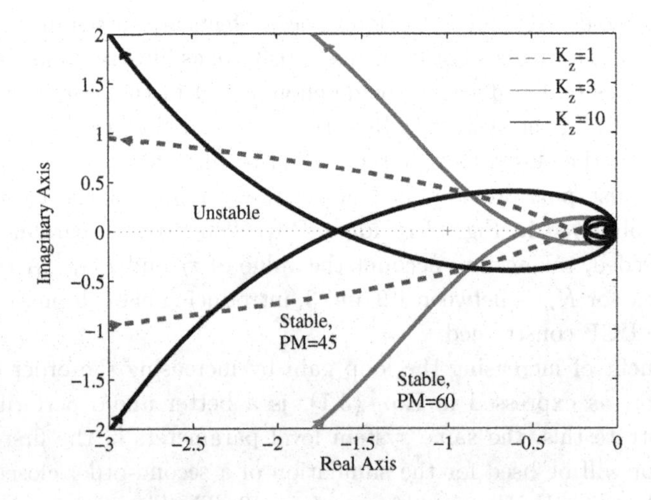

Fig. 3.10 Nyquist plot for second order PWM CDA for different K_z values.

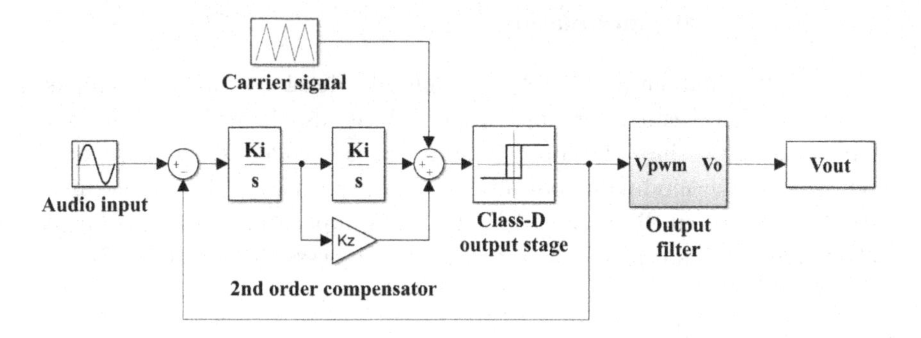

Fig. 3.11 Simulink model for 2nd order PWM CDA architecture.

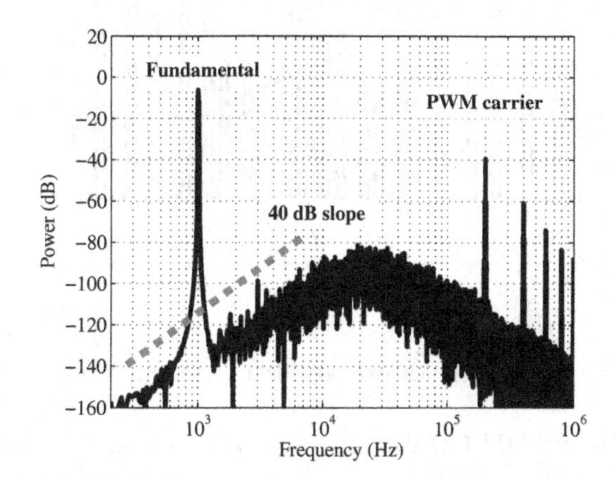

Fig. 3.12 Output spectrum for 2nd order PWM CDA with UGF = 50 kHz.

increasing the overall THD+N of the system compared to the first order system. Table 3.1 summarizes the simulation results for different compensator orders, UGF, or K_Z values. In conclusion, a high order compensator will help to attenuate the noise and distortion in the band of interest. However, the stability for the closed-loop system degrades as the compensator order increases. Higher order modulators have been used in closed-loop PWM CDA architectures to achieve high audio performance but at the expense of design complexity and power consumption [44, 72, 73, 75].

3.3 Sigma-delta modulation

Sigma-delta modulation (SDM) was initially developed for oversampled analog-to-digital converter applications. It is also known as pulse density modulation since the input information is encoded as the number of pulses that the modulator outputs, as observed in Fig. 3.13. The main difference between SDM and PWM for CDA applications is that a higher clock frequency is typically used to exploit the oversampling effect [77].

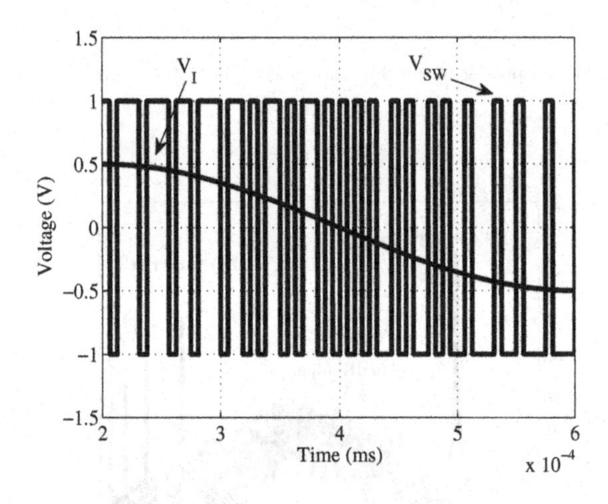

Fig. 3.13 Closed-loop SDM waveforms.

The closed-loop SDM CDA architecture is shown in Fig. 3.14, where a high frequency clock signal (V_{CLK}) is used to sample and hold the output of the compensator, and a quantization block is used to transform the voltage information to a pulse density encoding. The output of the quantizer is amplified in the class-D output stage and applied to the output filter.

Table 3.1 Summary for the closed-loop PWM CDA architecture simulations.

Order	UGF	DC gain	THD+N
1st	25 kHz	60 dB	−84 dB
1st	50 kHz	66 dB	−90 dB
1st	100 kHz	72 dB	−60 dB
2nd, K_z=1	50 kHz	100 dB	−92dB
2nd, K_z=2	50 kHz	120 dB	−98 dB
2nd, K_z=3	50 kHz	140 dB	−76 dB

Finally, the switching output of the CDA is used as the feedback signal and returned to the input of the compensator.

To understand the benefits of using SDM, the quantization error and the oversampling effect have to be detailed. The comparator typically used as quantizer in the CDA takes its input and compares it to a reference value (e.g. common mode voltage) to make a decision every rising edge of the clock signal. If the input signal is higher than the reference, the output changes to a high level; if the input signal is lower than the reference, the output changes to a low level. Thus, the difference between the input value and the output value of the quantizer is the quantization error. The number of quantization levels affects how large is the quantization error since each extra step is closer to the ideal value. The quantization step size (q) as a function of the number of bits in the quantizer is typically expressed as [84],

$$q = \frac{2 \cdot V_{DD}}{2^b - 1} \tag{3.6}$$

where V_{DD} is the supply voltage of the quantizer, and b represents the number of bits of the quantizer. Assuming uniform distribution of the quantization error (e_q) across all the quantization levels, we can express the average quantization error noise power as [84],

$$\sigma_e^2 = \int_{-q/2}^{q/2} \frac{e_q^2}{q} de_q = \frac{q^2}{12} = \frac{V_{DD}^2}{3.2^{2b}}. \tag{3.7}$$

The signal to noise ratio of an n-bit quantizer for a sinusoidal signal of amplitude A with signal power $\sigma_x^2 = A^2/2$ is expressed as,

$$\text{SNR (dB)} = 20 \cdot \log_{10} \frac{\sigma_x}{\sigma_e} \cong 20 \cdot \log_{10} \frac{A}{V_{DD}} + 6.02 \cdot b + 1.76. \tag{3.8}$$

It can be observed from Eq. (3.8) that increasing the number of bits, enhances the SNR of the system. For each additional bit in the quantizer,

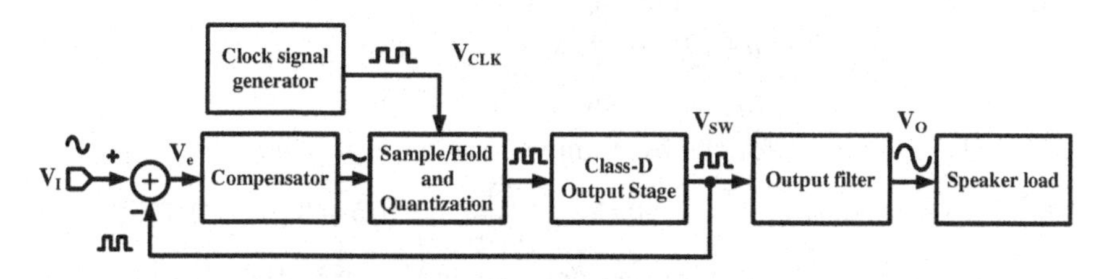

Fig. 3.14 Closed-loop SDM CDA architecture.

an extra 6 dB of SNR are added. However, more bits in the quantizer increase the complexity of the quantizer design and the power consumption of the overall system. To overcome this, oversampling can be used to decrease the quantization noise in the band of interest. The amount of oversampling is typically expressed as a function of the desired bandwidth as the oversampling ratio $OSR = 2^r = F_{SW}/(2 \cdot f_B)$ where r is the desired ratio as a power of 2 integer.

The oversampling can be better understood by observing its effect on the error power density as shown in Fig. 3.15. For a desired frequency bandwidth (f_B) sampled at the Nyquist rate of $F_{SW} = 2f_B$ with a power spectral density of $P_e(f)$ [84], the oversampling will spread the same $P_e(f)$ across the new bandwidth, lowering the in-band $P_e(f)$. Thus, if an ideal brick-wall filter is used at f_B, $P_e(f)$ will be reduced by the amount of oversampling used.

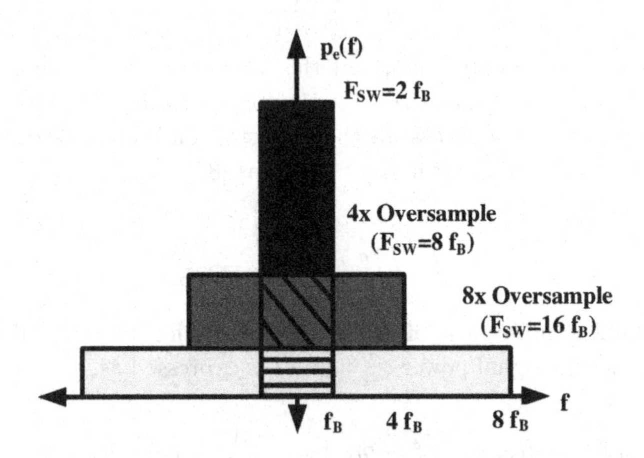

Fig. 3.15 Oversampling effect on the error power density.

The new in-band noise power after spreading the quantization error noise power (σ_e^2) with oversampling is,

$$\sigma_n^2 = \int_{-q/2}^{q/2} \frac{e_q^2}{q} \left(\frac{2 \cdot f_B}{F_{SW}} \right) de_q = \frac{q^2}{2^r \cdot 12} = \frac{\sigma_e^2}{OSR} \tag{3.9}$$

where the peak SNR for the oversampled system can be expressed as,

$$\text{SNR (dB)} = 20 \cdot \log_{10} \frac{\sigma_x}{\sigma_n} \cong 20 \cdot \log_{10} \frac{A}{V_{DD}} + 6.02b + 3.01r + 1.76. \tag{3.10}$$

It can be noticed that for every doubling in the OSR, the SNR improves by 3 dB. Thus, for low frequency applications such as audio, the OSR can be as high as 2^{10} improving the SNR by 30 dB only by using oversampling. However, this implies that a very high frequency clock signal has to be used, and that the circuitry needs to operate at higher frequency, increasing the power consumption of the system. The SDM architecture leverages the

Table 3.2 Ideal SNR for some SDM examples.

Order (N)	OSR	SNR (dB)
1	128	60
2	128	94
3	128	128
1	256	69
2	256	109
3	256	149

benefits of oversampling in a closed-loop system where a compensator helps to attenuate the distortion and quantization errors further, as in PWM architectures. The SDM can use a first order or higher order compensator to improve the CDA performance. However, the UGF of the loop is typically fixed to the audio bandwidth of 20 kHz. Thus, the OSR or the compensator order (N) are increased to improve the SNR. The Nth-order SDM peak SNR can be expressed as [84],

$$\text{SNR (dB)} \cong 6.02b+1.76+10\log_{10}(2N+1)-9.94N+3.01(2N+1)r. \quad (3.11)$$

It can be observed that the oversampling effect is enhanced by the compensator order by a factor of $(2N+1)$. Thus, high compensator order and high OSR can achieve outstanding performance. Table 3.2 summarizes a few examples for the calculation of the ideal SNR for a 1-bit quantizer SDM for several values of N and OSR. For high audio performance, the SDM CDA architecture would need a high OSR of at least 128 ($F_{SW} = 2 \cdot 128 \cdot 20$ kHz $= 5.12$ MHz), a high order compensator of at least 2nd order, or a combination of both.

A system level example using MATLAB© for a first-order SDM CDA was simulated for an input signal of 0.5 V_{RMS} at 1 kHz, $OSR = 128$ or $F_{SW} = 5.12$ MHz for a $f_B = 20$ kHz, and a Butterworth second order low pass output filter with cut-off frequency of 20 kHz. The Simulink model used for the simulation is illustrated in Fig. 3.16 where $Ki = 2 \cdot \pi \cdot f_B$. The frequency spectrum for the output signal is shown in Fig. 3.17.

It can be observed that the noise floor has a 20 dB/dec slope as expected. Also, the clock signal is at very high frequencies compared to the fundamental signal. Thus, the output filter attenuates it more, minimizing its effect on the speaker. The third harmonic is −80 dB from the fundamental tone, and the peak SNR is 60 dB.

To verify the effect of a higher compensator order as expressed in Eq. (3.11), N was increased to 2 with the same OSR = 128. The output frequency spectrum for this system is shown in Fig. 3.18. The second order compensator was implemented as in the PWM second order example shown in Fig. 3.11 but using SDM. It can be observed that the noise floor is attenuated with a 40 dB/decade slope, reducing the THD+N. By increasing the compensator order, the SNR improves as expressed in Eq. (3.11). Moreover, the higher compensator order, also helps to correct for distortion; the third harmonic is −86 dB from the fundamental tone, and the peak SNR is 72 dB. However, the SNR did not improve as predicted by

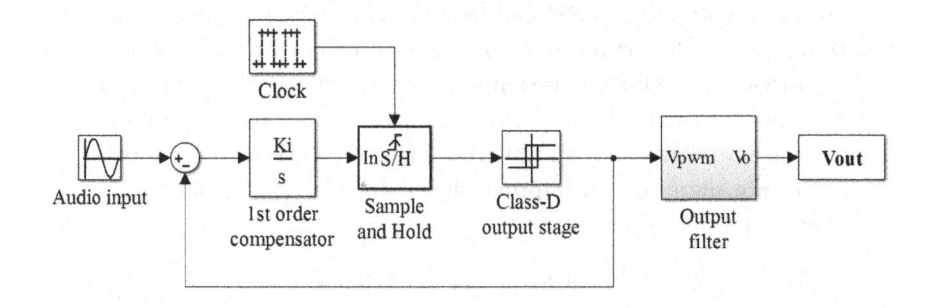

Fig. 3.16 Simulink model for 1st order SDM CDA architecture.

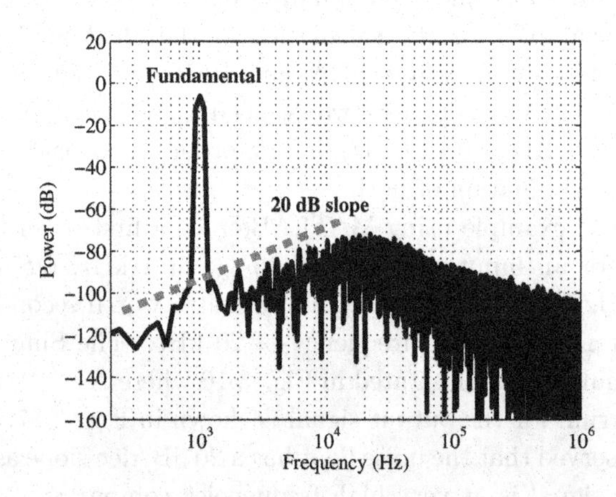

Fig. 3.17 Output spectrum for 1st order SDM CDA with OSR = 128.

Eq. (3.11) since even in an ideal simulation, the SDM is affected by timing errors in the simulator solver engine, the choice of the compensation zero frequency location, and other implementation inaccuracies.

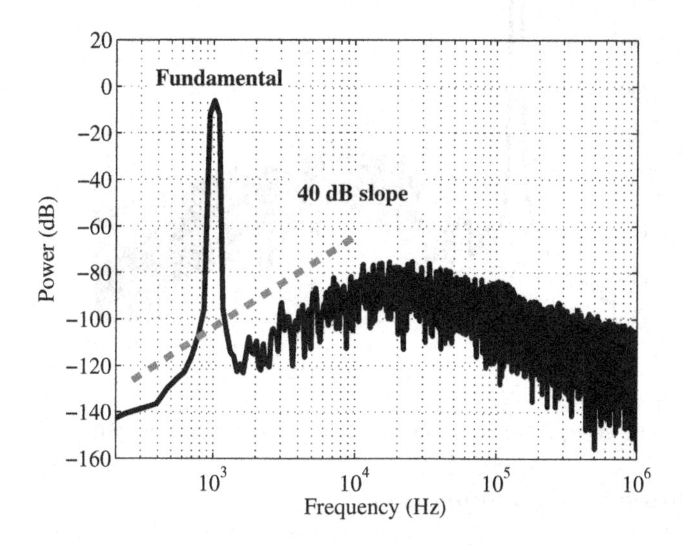

Fig. 3.18 Output spectrum for 2nd order SDM CDA with OSR = 128.

It can be observed that increasing the OSR also improves the performance of the system as expressed in Eq. (3.11). To prove this, the OSR was increased to 256 for the second order SDM CDA, and the output frequency spectrum is shown in Fig. 3.19. It can be observed that the same 40 dB/decade slope in the noise floor remains but the noise and harmonics are attenuated even more due to the higher OSR. The third harmonic is -100 dB from the fundamental tone, and the peak SNR is 86 dB. It is worth noticing that the SDM has also been proposed for DC-DC buck converters [85], where the main goal of using SDM is to reduce both the carrier distortion at the output and the EMI produced by the converter itself.

In conclusion, the SDM gives the advantage of using a high frequency clock signal to perform the oversampling instead of a triangle waveform as in PWM. The audio performance is highly dependent on the choice of compensator order and OSR; as the OSR increases, the circuit requirements on each block are more demanding. Also, as the compensator order increases, the stability of the loop starts to degrade, requiring careful choice of compensation.

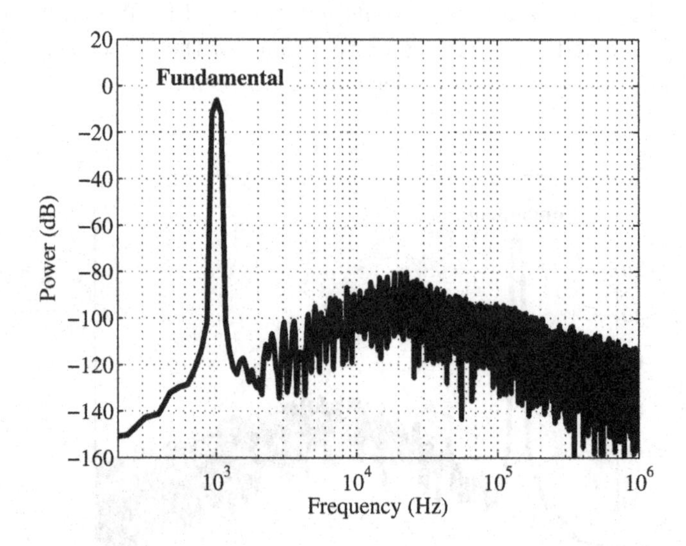

Fig. 3.19 Output spectrum for 2nd order SDM CDA with OSR = 256.

3.4 Self-oscillating modulation

The self-oscillating modulation (SOM) is inherently a closed-loop architecture that leverages the fact that the compensator introduces a signal delay to implement an oscillator with oscillating frequency at F_{SW}. This architecture meets the Barkhausen criteria (BKC) for the closed-loop system in Eq. (3.1) expressed as,

$$|LG(s)| = |A_{ol}(s) \cdot \beta(s)| = 1, \tag{3.12}$$

$$\angle LG(s) = \angle(A_{ol}(s) \cdot \beta(s)) = 2\pi n, n \in 0, 1, 2, \ldots \tag{3.13}$$

where the loop gain magnitude of the system must be 1 and the phase shift of the loop gain must be a multiple of 2π (e.g. $0°, 360°, \ldots$). The general SOM CDA architecture is shown in Fig. 3.20, where $LG(s)$ is determined by the compensator transfer function, the hysteretic comparator, and the output stage transfer function. The hysteretic comparator plays an important role in the operation of the SOM since it changes its magnitude and phase shift proportionally to the input signal [70].

The input signal dependency of the hysteretic comparator [86] can be observed from its describing function evaluation for a sinusoidal input signal

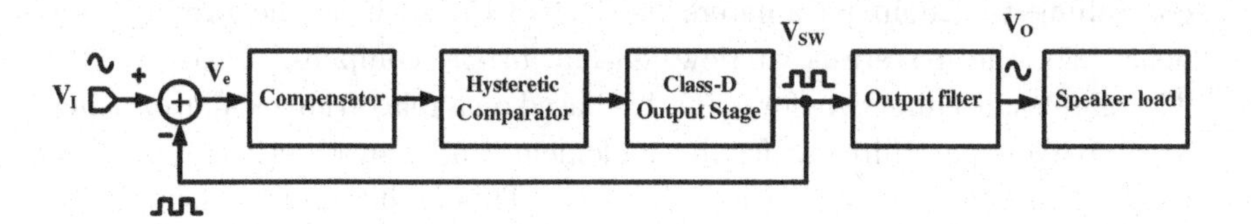

Fig. 3.20 Closed-loop general SOM CDA architecture.

with amplitude V_I that can be expressed as,

$$G_M(V_I) \cong \frac{V_{DD}}{V_I} e^{-j \sin^{-1}(V_H/V_I)} \tag{3.14}$$

where V_H is the hysteresis window, V_{DD} is the voltage limit or supply of the comparator, and assuming that $V_I > V_H$. The BKC for the SOM system must be accomplished by the negative feedback in the loop that adds $-180°$ of phase shift while the compensator ideally adds less than $-180°$ of phase shift to remain stable. Therefore, to satisfy Eq. (3.13), the loop changes the hysteretic comparator phase shift by manipulating the amplitude of its input signal (V_I). Moreover, this change in V_I also changes its magnitude gain to remain within the hysteresis window such that the overall loop gain satisfies Eq. (3.12). The main advantage of this architecture is that it obviates the need of a clock or carrier signal generator; since it generates its own switching signal, the SOM saves power and reduces the complexity of the loop. The main drawback is that its switching frequency could decrease too much for large input amplitudes, degrading the THD+N. The comparator hysteresis window, the compensator delay, and the propagation delay in the loop will determine the modulation frequency of the SOM system [70]. The F_{SW} as a function of duty cycle ($D = V_I/V_{supply}$) could be expressed as,

$$F_{SW}(D) = \frac{D \cdot (1 - D)}{\frac{V_H \cdot \tau_{compensator}}{V_{supply}} + \tau_d} \tag{3.15}$$

where V_{supply} is the supply voltage of the comparator, V_H is the voltage hysteresis window of the comparator, $\tau_{compensator}$ is the compensator delay, and τ_d is the propagation delay from the comparator to the input of the compensator, including the comparator delay, the output stage delay, and the feedback network delay.

It can be noticed that F_{SW} is not constant since it chances as a function of the input signal. The F_{SW} variation could be reduced if needed by controlling the main parameters in Eq. (3.15) such as the propagation delay [87], the hysteresis window [88, 89], or the compensator delay [90]. The simplest architecture for a SOM is based on a relaxation oscillator using a passive low pass filter as feedback element, and a hysteretic comparator to close the loop, as shown in Fig. 3.21. This architecture is commonly known as bang-bang (BB) architecture since the output only changes when the error signal is higher/lower than the hysteresis window [69].

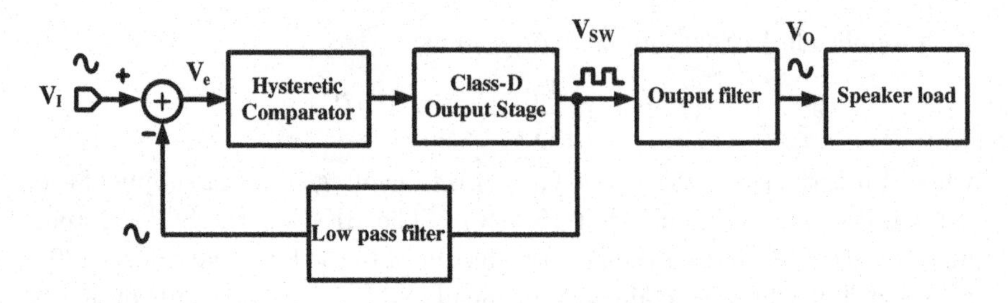

Fig. 3.21 Bang-bang SOM CDA architecture.

A system level simulation using MATLAB © was performed for a bang-bang CDA to verify its functionality with a hysteresis window of 0.1 V, supply voltage of 1 V, and low pass filter pole at 20 kHz ($wps = 2 \cdot \pi \cdot 20$ kHz), as illustrated in the Simulink model of Fig. 3.22. The output power spectrum is shown in Fig. 3.23. It can be observed that the SOM carrier frequency is not a pure tone and that its power is spread across a frequency range proportional to the amplitude of the input as expected from Eq. (3.15). Also, the loop magnitude gain is only determined by the hysteretic comparator, as expressed in Eq. (3.14), that could not be enough to achieve high performance; the third harmonic is at −64 dB from the fundamental harmonic.

The simplicity and small size of the bang-bang CDA comes at the expense of limited linearity performance due the absence of a compensator to provide gain in the loop. Its performance can only be improved by reducing the hysteresis window or by changing the low pass filter pole location to increase F_{SW} as expressed in Eq. (3.15).

A high order compensator in the general SOM CDA architecture of Fig. 3.20 will provide better audio performance due to the increased loop gain at the expense of increased power consumption and design complexity [70, 83, 90, 91]. A system level simulation of a second order general SOM CDA architecture was performed with the same parameters as the bang-bang

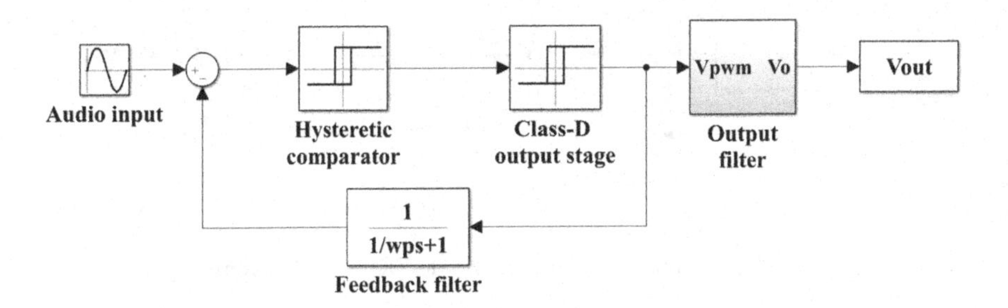

Fig. 3.22 Simulink model for bang-bang CDA architecture.

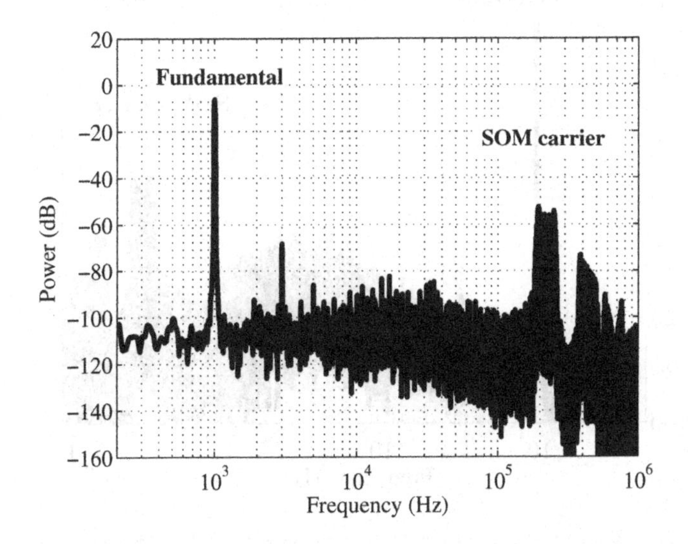

Fig. 3.23 Output spectrum for bang-bang CDA architecture.

CDA architecture, as illustrated in the Simulink model in Fig. 3.24 where $Ki = wps = 2 \cdot \pi \cdot 20$ kHz, and $K_Z = 1.5$. The output frequency spectrum is shown in Fig. 3.25. It can be observed that the noise floor is highly attenuated due to the large loop gain contributed by the compensator, and it has a 40 dB/decade slope due to the double integrator. The third harmonic is at -110 dB from the fundamental harmonic. Also, the F_{SW} changed since the compensator delay decreased. Nonetheless, the same power spreading of the SOM carrier happens. Differently from the general

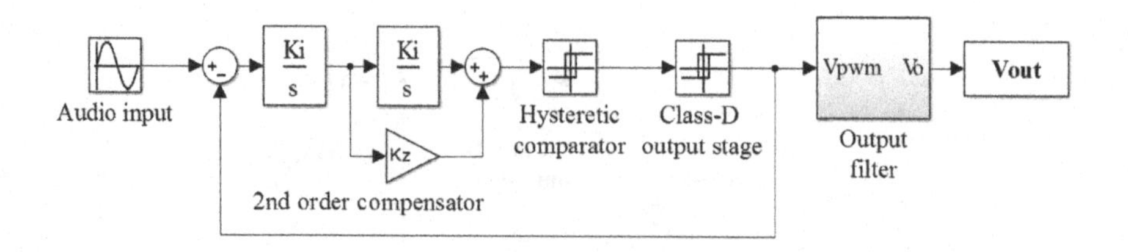

Fig. 3.24 Simulink model for a 2nd order SOM CDA architecture.

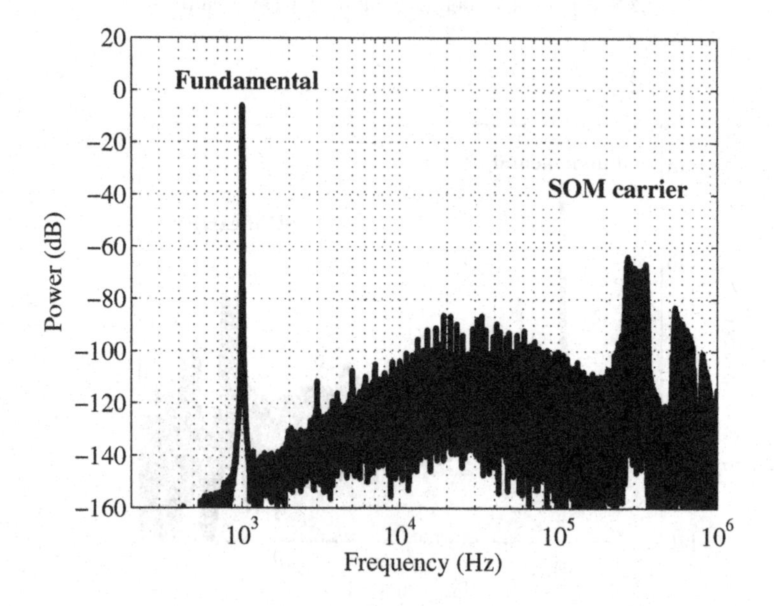

Fig. 3.25 Output spectrum for 2nd order general SOM CDA architecture.

architecture of Fig. 3.20, other SOM systems include the output filter in the loop and use a non-linear control technique as a compensator to remain stable, as shown in Fig. 3.26. This non-linear control technique is known as sliding mode control (SMC), and it is implemented as a tracking system with robust operation under external perturbations [92].

The SMC architecture provides the advantage of relaxed design requirements in the compensator since the feedback signal includes low frequency signal components, and the output filter errors are attenuated by the loop gain of the system [92]. The non-linear controller ensures stability in the CDA with high audio performance and low power consumption [78–81, 93].

The SMC is based on the state variables of the switching structure of the system. For the CDA, the state variables are the inductor current (i_L) and capacitor voltage (v_C) in the output low pass filter. The CDA has two different structures, as observed in Fig. 3.27, depending on the switching state of the output.

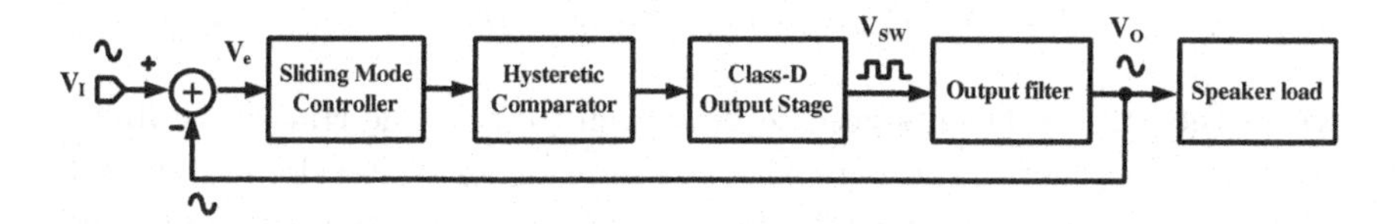

Fig. 3.26 Closed-loop SMC CDA architecture.

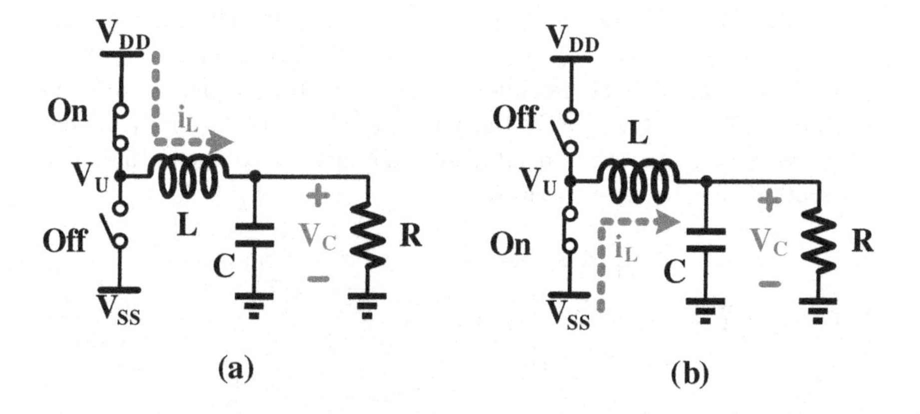

Fig. 3.27 Switching structure in the CDA when connected to (a) V_{DD} or (b) V_{SS}.

The state-space model corresponding to the output filter of the CDA, as shown in Fig. 3.27, can be expressed as,

$$\frac{d}{dt}\begin{pmatrix} i_L(t) \\ v_C(t) \end{pmatrix} = \begin{pmatrix} 0 & -\dfrac{1}{L} \\ \dfrac{1}{C} & -\dfrac{1}{CR} \end{pmatrix} \begin{pmatrix} i_L(t) \\ v_C(t) \end{pmatrix} + \begin{pmatrix} \dfrac{1}{L} \\ 0 \end{pmatrix} V_U(t), \tag{3.16}$$

where $v_C(t)$ is the voltage across the capacitor C, $i_L(t)$ is the current through the inductor L, R represents the speaker resistance, and $V_U(t)$ is the binary-modulated signal generated by the SMC. The goal of the SMC is to generate the control signal $V_U(t)$ using a control law defined by a switching function to force the system to follow a desired response according to the system sliding equilibrium point [94].

The switching function in the SMC is typically chosen to minimize the error voltage of the system, $V_e(t) = V_I(t) - V_O(t)$, and it is defined in the canonical form for a second order system as,

$$s(V_e, t) = k_1 V_e(t) + k_2 \dot{V}_e(t) \tag{3.17}$$

where the kth coefficients need to be chosen to meet the Hurwitz stability criterion. In general, the switching function for an Nth-order system will be an $(N-1)$ order system [92]. Thus, $N = 1$ is needed to control a second order system which simplifies the compensator design requirements. Also, the switching function in Eq. (3.17) implies that a differentiation is needed which does not attenuate the in-band noise [78, 82].

A system level simulation was performed for the SMC CDA architecture with $k_1 = 1$ and $k_2 = \frac{1}{2\pi \cdot 28\ \text{kHz}}$ [78, 82], and the output power spectrum using the Simulink model illustrated in Fig. 3.28. The frequency spectrum is shown in Fig. 3.29, where the third harmonic is -80 dB from the fundamental harmonic, but the in-band noise is not attenuated due to the use of differentiators in the compensator.

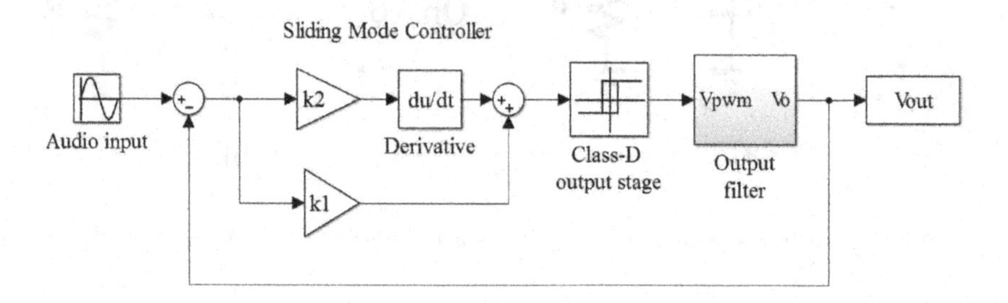

Fig. 3.28 Simulink model for SMC CDA architecture.

Another alternative is to redefine the switching function in Eq. (3.17) using integrals instead of differentiations [79, 81]. By doing this, the compensator attenuates the in-band noise, and has more relaxed design requirements. This alternative is known as integral sliding mode control (ISMC) because it uses integrators instead of differentiators, and its architecture is shown in Fig. 3.30. Note that an additional feedback signal is added before the hysteretic comparator, using the inductor current information to reduce the state-space system to a first order. Thus, the new switching

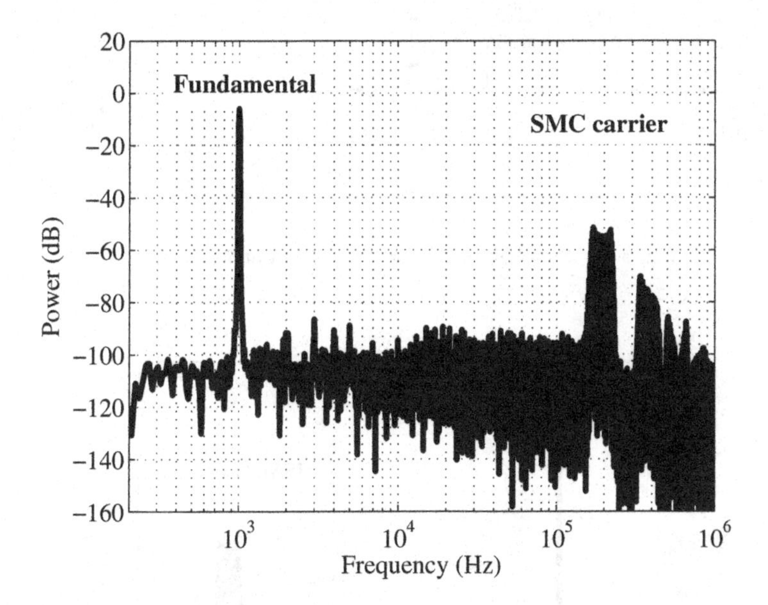

Fig. 3.29 Output spectrum for SMC CDA architecture.

function for the ISMC [79] is defined as,

$$s(V_e, t) = k_i \int V_e(t) - i_L(t). \tag{3.18}$$

The drawback of the ISMC architecture is that the inductor current information, containing high frequency components at F_{SW}, needs to be sensed; thus, the circuit that implements the adder before the hysteretic comparator has to be able to process the high frequency signals with accuracy, increasing its power consumption.

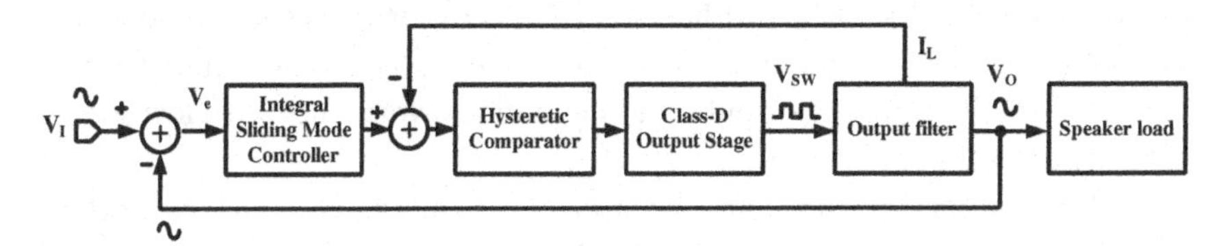

Fig. 3.30 Closed-loop ISMC CDA architecture.

A system level simulation was performed for the ISMC CDA architecture with $k_i = 2\pi \cdot 20$ kHz [79, 81] using the Simulink model illustrated

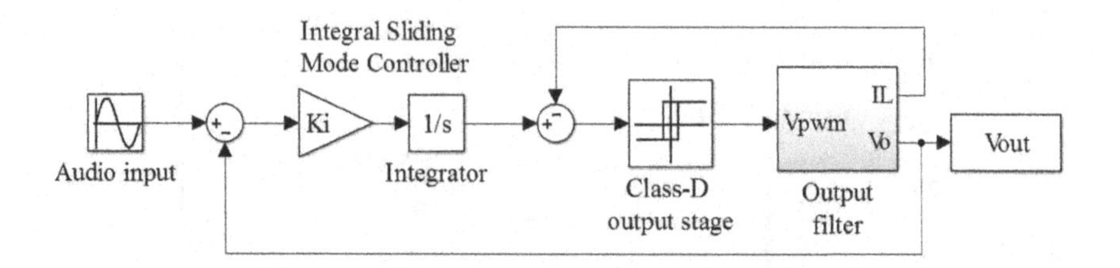

Fig. 3.31 Simulink model for ISMC CDA architecture.

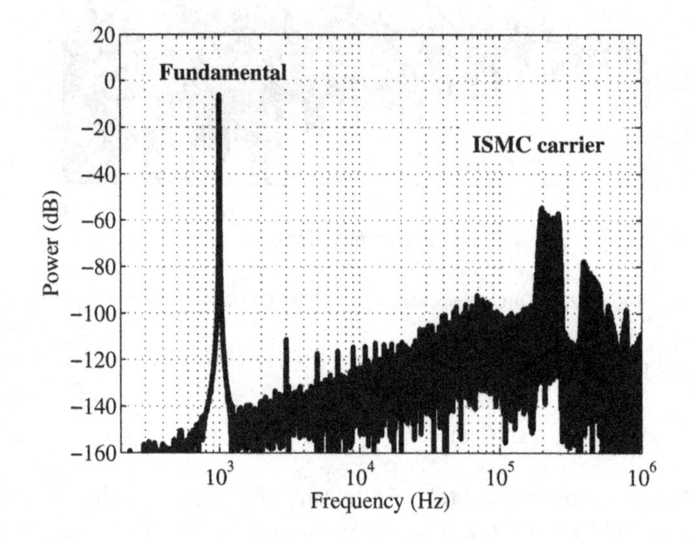

Fig. 3.32 Output spectrum for ISMC CDA architecture.

in Fig. 3.31, and the output power spectrum is shown in Fig. 3.32. It can be observed that the third harmonic is −110 dB from the fundamental harmonic, and the in-band noise is greatly attenuated by the compensator. Thus, the ISMC can provide outstanding audio performance with low power consumption.

SOM architectures have also been proposed in DC-DC buck converters where the inherent stability and fast transient response is leveraged for integrated power management modules [87, 95, 96]. In conclusion, the SOM CDA architecture obviates the need for a carrier signal generator, and it provides high audio performance, low power consumption, and high efficiency. The tradeoffs are its variable switching frequency, but this can be addressed using different schemes to keep a quasi-constant F_{SW}. Detailed information and design procedures of CDAs using SOM, SMC, and ISMC will be addressed in later chapters of this book.

3.5 Comparison between modulation schemes

To summarize the advantages and tradeoffs in the main closed-loop CDA architectures in terms of audio performance, Table 3.3 presents the simulation results for the main architectures and some estimations about the circuit design complexity, EMI, power, and area.

Table 3.3 Closed-loop CDA architecture audio performance comparison.

System	Order	THD+N	SNR	Complexity	EMI	Power	Area
PWM	1	−76 dB	84 dB	L	H	L	M
PWM	2	−92 dB	90 dB	M	H	M	M
SDM	1	−81 dB	86 dB	M	L	M	M
SDM	2	−88 dB	92 dB	H	L	H	H
SOM	1	−90 dB	86 dB	M	M	L	L
SOM	2	−99 dB	100 dB	M	M	M	M
BB	1	−62 dB	86 dB	L	M	L	L
SMC	1	−80 dB	90 dB	M	M	L	M
ISMC	1	−104 dB	120 dB	L	M	L	L

L = Low, M = Medium, H = High

It can be observed that all the closed-loop CDA architectures could provide good audio performance, with THD+N < -60 dB and SNR > 80 dB. The preferred choice for a CDA architecture in commercial applications is the PWM due to its medium complexity, area, and power tradeoffs [44,56]. The SDM architecture is typically not chosen due to its relative high complexity, power, and area consumption [77], but still provides a good alternative for digital input CDA due to its compatibility with sampled data. The SOM architectures provide a good alternative for high performance CDA, but their development have remained as academic research [78–81,83], and their commercial applications are limited. Note that the PWM carrier has tones at a fixed frequency, while SDM and SOM carriers have tones spread over the whole Nyquist band. This can lead to crosstalk problems with other surrounding blocks and IMD at unknown frequencies; whereas with PWM, the crosstalk and IMD frequencies are known and can be taken into account [44].

Chapter 4

Class-D Circuit Design Techniques

4.1 System implementation

In general, the ideal system level results in Table 3.3 will be degraded by the non-ideal circuit implementation of each building block in the architecture. The non-idealities in the amplifiers used in the compensator, the timing errors in the class-D output stage, the output filter non-idealities, among other perturbations will limit the maximum achievable THD+N and SNR in the system.

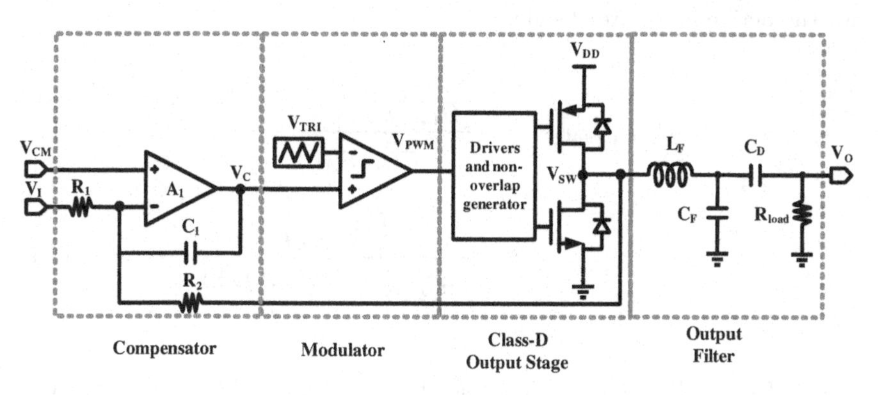

Fig. 4.1 First-order single-ended PWM class-D audio amplifier circuit.

To illustrate the circuit design tradeoffs, a conventional single-ended (SE) first-order PWM CDA architecture is built as shown in Fig. 4.1. The compensator is implemented using an amplifier (A_1) configured as an active integrator. The PWM modulator is chosen as an open-loop comparator with a high frequency triangle wave as the carrier signal. The class-D output stage is designed using a CMOS switch with drivers and non-overlap generator circuits. The output filter is built as a second-order LC low-pass filter with AC coupling to remove the DC component at the speaker. The circuit design tradeoffs for each block will be detailed next, and an example implementation in a standard 0.18 μm CMOS technology will be provided.

4.2 Compensator design

The compensator design is important because it limits the output noise and distortion performance of the system. It is typically implemented as a chain of integrators, where each integrator is realized using an amplifier in feedback. An ideal amplifier would provide infinite DC gain and unlimited gain bandwidth product (GBW); but, in reality, the amplifier will have a finite gain and limited GBW that will degrade the implemented function. To achieve high audio performance in the CDA, it is necessary to understand the effects of the non-ideal amplifier in the integrator function.

The active integrator ideal time constant is $\tau_I = R_1 \cdot C_1$, and its value selection depends on several tradeoffs in the CDA, between the passive component values, amplifier A_1 power consumption, linearity, and in-band noise. Considering the finite gain of the amplifier A_1, the transfer function for the active integrator yields,

$$
\begin{aligned}
G_{int,actual}(s) &= -\frac{G_{int,ideal}(s)}{1+\frac{1}{A_1(s)\cdot\beta_C(s)}} \\
&\cong -\frac{\left(\frac{1}{s\cdot R_1\cdot C_1}\right)}{1+\frac{s}{GBW}\cdot\left(\frac{s\cdot R_{1,2}\cdot C_1+1}{s\cdot R_{1,2}\cdot C_1}\right)}
\end{aligned}
\tag{4.1}
$$

where $R_{1,2} = R_1//R_2$, the amplifier transfer function is characterized as $A_1(s) \cong GBW/s$, and $\beta_C(s)$ is the integrator amplifier feedback transfer function.

The magnitude and phase of Eq. (4.1) for $GBW \cdot R_{1,2} \cdot C_1 \gg 1$ can be expressed as,

$$G_{int,actual}(j\omega) \cong -\frac{1}{\frac{j\omega}{\omega_{int}} - \frac{\omega^2}{GBW \cdot \omega_{int}}},$$

$$|G_{int,actual}(j\omega_{int})| \cong \frac{1}{\sqrt{1+\left(\frac{\omega_{int}}{GBW}\right)^2}}, \quad (4.2)$$

$$\angle G_{int,actual}(j\omega_{int}) \cong -90^\circ - \tan^{-1}\left(\frac{\omega_{int}}{GBW}\right)$$

where $\omega_{int} = 1/R_1 \cdot C_1$. The ideal integrator at ω_{int} will have a magnitude of one and a phase shift of -90°. Thus, the magnitude (Δ_M) and phase (Δ_φ) deviations for the non-ideal integrator for $GBW \gg \omega_{int}$ can be derived from Eq. (4.2) as,

$$\Delta_M = \frac{1-\sqrt{1+\left(\frac{\omega_{int}}{GBW}\right)^2}}{\sqrt{1+\left(\frac{\omega_{int}}{GBW}\right)^2}} \cong \frac{1}{2} \cdot \left(\frac{\omega_{int}}{GBW}\right)^2, \quad (4.3)$$

$$\Delta_\varphi = -\tan^{-1}\left(\frac{\omega_{int}}{GBW}\right).$$

One typical example is $GBW = 10 \cdot \omega_{int}$ that will give $\Delta_M \cong 0.5\%$ and $\Delta_\varphi < 6^\circ$. Another way to quantify the active integrator performance is in terms of its quality factor (Q_{int}), considering it as a reactive element $X(j\omega)$ with a lossy resistive part Re. The higher the value of Q_{int}, the better the integrator. From Eq. (4.2), Q_{int} for the active integrator implementation can be expressed as,

$$Q_{int} = \frac{X(j\omega)}{Re} \cong \frac{\omega \cdot R_1 \cdot C_1}{\frac{-\omega^2 R_1 \cdot C_1}{GBW}} = -\frac{GBW}{\omega} = -|A_1(j\omega)|. \quad (4.4)$$

The goal of the implementation is to minimize Δ_M and Δ_φ, avoiding significant deviations in the integrator performance. Thus, the amplifier GBW has to be higher than the chosen ω_{int}. The UGF of the closed loop first-order PWM CDA is given by $UGF(Hz) = \omega_{int}/(2\pi)$. Thus, a large value of ω_{int} would provide a high bandwidth for the CDA loop that will result in lower THD+N, higher PSRR at high frequencies, and smaller values for the passive components. However, the amplifier power consumption would need to be increased to avoid deviations in ω_{int} due to finite GBW. On the other hand, a small value of ω_{int} would result in higher THD+N, lower PSRR at high frequencies, and larger values for the passive components. However, the amplifier design requirements are relaxed, leading to low power consumption.

A good tradeoff between performance and power consumption for the unity-gain frequency (UGF) of the first-order PWM CDA architecture is 1/10 of F_{SW}; thus, for this example implementation, UGF= 50 kHz is selected for the system with F_{SW}= 500 kHz. The passive component selection to implement τ_I needs to consider the input resistors (R_1, R_2 in Fig. 4.1) noise contribution and the amplifier driving capability. The resistor thermal noise for 1 Hz bandwidth can be expressed as [61],

$$V_R^2(f) = 4kTR \cong 4kTR_s \frac{L_R}{W_R} + R_C \tag{4.5}$$

where $k = 1.38\text{x}10^{-23}$ is the Boltzmann constant, T is the temperature in Kelvins, R_s is the resistance per square for the chosen resistor type, L_R is the length of the resistor layout, W_R is the width of the resistor layout, and R_C is the total resistance due to all the contacts in the resistor layout. Thus, a large resistor value will introduce large thermal noise that can limit the THD+N and SNR performance of the whole system; but, a large capacitor will load the amplifier, requiring large power consumption to drive it. Also, depending on the capacitor implementation available in the technology, the capacitor could occupy more silicon area than the resistor; thus, the resistor design has a tradeoff between area, noise, and power consumption.

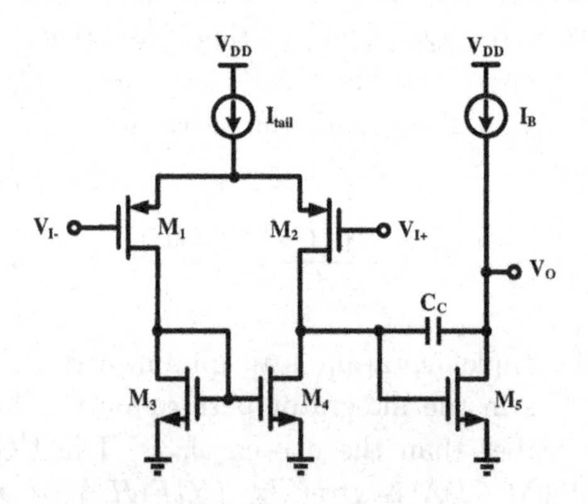

Fig. 4.2 Miller compensated two-stage amplifier.

The amplifier A_1 is often implemented as a two-stage Miller-compensated amplifier as shown in Fig. 4.2, where the Miller compensation capacitor (C_C) implements the pole-splitting technique to ensure a stable system [61,97–99]. The main design parameters for the amplifier are its input referred noise, slew rate (SR), DC gain (A_{DC}), and GBW. Each design parameter has its own tradeoffs which are actually correlated.

From a design perspective, the only variables in the amplifier design are the tail current (I_{tail}), the dimensions of each transistor, the bias current of the second stage (I_B), and any other passive component that is used for compensation. In general, the performance of most amplifier topologies are greatly dependent on the transconductance (g_m) of the input transistors (M_1, M_2). A large g_m in the input pair will reduce the thermal noise, provide large DC gain, and increase the GBW, but at the expense of increased power consumption as will be detailed next.

The design equations for g_m as a function of the transistor dimensions and biasing operating point can be expressed as,

$$g_m = \frac{I_{tail}}{V_{dsat}} = \frac{W}{L}\mu_p C_{ox} V_{dsat} = \sqrt{\frac{W}{L}\mu_p C_{ox} I_{tail}} \tag{4.6}$$

where μ_p is the mobility parameter for PMOS, W is the transistor width, L is the transistor length, C_{ox} is the oxide capacitance, and V_{dsat} is the voltage difference between the gate to source voltage (V_{gs}) and the threshold voltage (V_{th}) of the transistor defined as,

$$V_{dsat} = V_{gs} - V_{th} = \sqrt{\frac{I_{tail}}{\frac{W}{L}\mu_p C_{ox}}}. \tag{4.7}$$

It can be observed, that the g_m can be increased by having a large I_{tail}, a large W/L ratio, or by increasing both parameters. However, increasing I_{tail} will also increase the power consumption of the amplifier, and increasing the W/L ratio will reduce V_{dsat} too much, making the amplifier sensitive to process variations [61, 100]. Note that increasing W/L ratio too much leads the transistor to operate in weak inversion or subthreshold region, making g_m independent of W/L from there on.

The equivalent input referred noise for low and moderate frequencies in a MOS transistor is typically expressed as [61],

$$V_n^2(f) = V_{thermal}^2 + V_{flicker}^2 = 4kT\left(\frac{2}{3}\right)\frac{1}{g_m} + \frac{K_f}{WLC_{ox}f} \tag{4.8}$$

where K_f is a device-dependent parameter. Assuming two uncorrelated noise sources, it can be observed that to reduce the thermal noise contribution from the amplifier, a large g_m is needed, but this will require a large

I_{tail} and/or large W in the transistors. It is important to notice that the noise of transistors $M_3 - M_5$ is attenuated by the g_m of the input transistor, minimizing their contribution to the equivalent input noise. Thus, the input transistor noise contribution is critical and must be minimized. To reduce their flicker noise contribution, large W and L are needed but at the expense of increased active area. The input transistors (M_1, M_2 in Fig. 4.2) in A_1 are frequently chosen as PMOS transistors since the parameter K_f is smaller than in the NMOS. Also, wide L is typically used to minimize their flicker noise contribution in the audio bandwidth.

The GBW and A_{DC} for the amplifier topology shown in Fig. 4.2 are,

$$GBW = \frac{g_{m1,2}}{2\pi \cdot C_C}, \tag{4.9}$$

$$A_{DC} = \frac{g_{m2}}{g_{ds4}}\left(\frac{g_{m5}}{g_{ds5}}\right) \tag{4.10}$$

where g_{ds} is the drain to source conductance of the transistor, and $M_1 = M_2$ and $M_3 = M_4$. Typically, g_m is chosen as a function of the desired GBW as expressed in Eq. (4.9). Then, the g_{ds} of the transistors are designed to satisfy the required DC gain. To choose the amplifier GBW, several tradeoffs need to be considered such as SR, power consumption, and its effect on Eq. (4.1).

The amplifier slew rate imposes a limitation in the large signal operation of the compensator. In the two-stage miller-compensated amplifier topology, the worst case SR for a unity gain configuration can be expressed as [61, 100, 101],

$$SR \cong \frac{I_{tail}}{C_C} = \frac{I_{tail} \cdot 2\pi \cdot GBW}{gm_{1,2}} = 2\pi \cdot GBW \cdot V_{dsat_{1,2}}. \tag{4.11}$$

The full-power bandwidth is defined as the maximum frequency (f_{max}) at which the amplifier will yield an undistorted AC output with the largest possible amplitude (V_{max}) [100]. The minimum SR requirement for amplifier A_1 using this definition is,

$$SR_{min} \geq 2\pi \cdot f_{max} \cdot V_{max}. \tag{4.12}$$

The compensator amplifier A_1 has to process the input audio signal and the high frequency feedback signal. Therefore, f_{max} should correspond to the feedback switching frequency, and V_{max} to the peak voltage of the error signal in the system. To specify a minimum GBW requirement, the small signal behavior in Eq. (4.11) could be related to the large signal by using

Eq. (4.12), and solving for GBW [101]. The minimum GBW needed in amplifier A_1 can be expressed as,

$$GBW_{min} \geq \frac{f_{max} \cdot V_{max}}{V_{dsat_{1,2}}}. \tag{4.13}$$

Several design alternatives are possible taking into account the tradeoffs present in Eq. (4.13). Also, the $V_{dsat_{1,2}}$ design choice presents tradeoffs in the amplifier A_1 between its DC gain, offset voltage, noise, bandwidth, and stability [61, 100].

Table 4.1 Design procedure for active integrator.

1. Choose R_1, R_2 values based on the desired output noise.
2. Determine $C_1 = 1/(\omega_{int} \cdot R_1)$.
3. Find minimum GBW to satisfy Eq. (4.13).
4. Select C_C to ensure stability for the obtained GBW_{min}.
5. Calculate minimum I_{tail} to satisfy Eq. (4.11).
6. Find required $g_{m1,2}$ using Eq. (4.9).
7. Select input transistor width to meet Eq. (4.6), using a wide length value to lower noise contribution.
8. Design remaining transistors $M_3 - M_5$ to ensure high DC gain, using Eq. (4.10).

Table 4.1 summarizes the design procedure for the active integrator used in the compensator to determine the component values of the amplifier given the output noise, V_{max}, f_{max}, $V_{dsat1,2}$, and ω_{int}. For this example, the input resistors R_1, R_2 were chosen as 400 kΩ and the integrating capacitor C_1 as 8 pF. This will result in an integrated noise in the 20 kHz bandwidth of 11.37μV at 300° K. The amplifier A_1 uses $V_{dsat_{1,2}} = 100\ mV$ for a $f_{max} = 500$ kHz with $V_{max} = 0.9\ V$, which would require a minimum GBW of 4.5 MHz. Since the GBW parameter is chosen to satisfy Eq. (4.13), A_1 DC gain is selected taking into account the tradeoffs between the integrator performance and the amplifier power and area consumption, as detailed in Eq. (4.6) to Eq. (4.10). For an integrating capacitor (C_1 in Fig. 4.1) of 8 pF, C_C for phase margin of 60° is typically chosen as $C_C > 0.22C_1$ [61]. Thus, C_C= 2pF is chosen. The required g_m arising from the C_C and GBW selection is 60 μS, requiring $I_{tail} = 6$ μA for $V_{dsat} = 0.1\ V$, as expressed in Eq. (4.6).

The remaining transistors $M_3 - M_5$ in A_1 are designed to maximize their conductance to obtain high DC gain, as expressed in Eq. (4.10). The simulated frequency response of the designed amplifier is shown in Fig. 4.3, where the DC gain and GBW requirements are satisfied, and stability is achieved with PM of 60°. This design selection yields a magnitude and phase deviation in the integrator function of 0.0025% and 0.4° as expressed in Eq. (4.3), respectively.

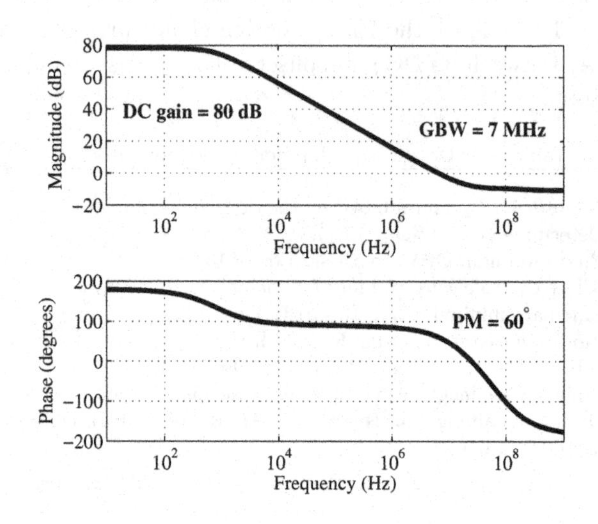

Fig. 4.3 Bode plot of two-stage example amplifier.

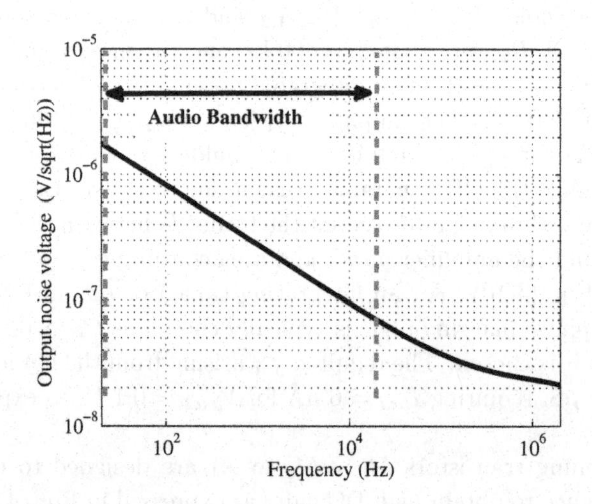

Fig. 4.4 Equivalent output noise voltage for the example amplifier.

The equivalent output noise voltage for the designed amplifier example is shown in Fig. 4.4. The integrated noise in the audio bandwidth is 13 μV at 300° K, that is comparable to the input resistor noise. This is important since these two noise sources will dominate the output noise of the CDA. If the resistor noise is higher than the amplifier noise, then the amplifier power consumption can be lowered to have less or equal noise than the input resistors. On the other hand, if the resistor noise is lower than the amplifier noise, then the amplifier power consumption has to be increased to have less or equal noise than the input resistors, or to drive a larger capacitance since the resistor value is decreased to lower its noise.

The total amplifier power consumption from a 1.8 V supply (V_{DD}) is 56 μW with an I_{tail} of 10μA, where the first stage consumes 18 μW and the second stage consumes 38 μW. The second stage often consumes more power than the first stage to push the output pole to high frequencies, ensuring a single pole frequency response in the amplifier [61, 100].

4.3 Pulse-width modulator

The PWM circuit can be implemented using different comparator structures to achieve large DC gain [61, 100, 102]. One option is to design the comparator as a push-pull open loop amplifier, as shown in Fig. 4.5, to achieve high slew rate to minimize propagation delay and high PSRR to suppress the supply noise contribution at its output. This implementation is also known as the three current mirror transconductance amplifier.

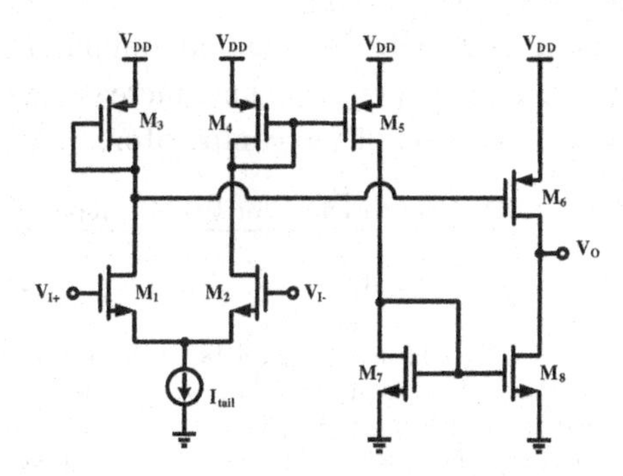

Fig. 4.5 Comparator schematic diagram with push-pull output stage.

The comparator takes the input voltage difference and amplifies it such that the output is pushed up to the supply voltage or pulled down to ground.

The DC gain of the comparator is given by,

$$A_{DC,comp} = \frac{g_{m1}}{g_{m3}}\left[\frac{(W/L)_6}{(W/L)_3}\right] \cong \frac{(W/L)_1}{(W/L)_3}\left[\frac{(W/L)_6}{(W/L)_3}\right] \tag{4.14}$$

where transistors are assumed as $M_1 = M_2$, $M_3 = M_4$, $M_5 = M_6$, and $M_7 = M_8$. The main advantage of this implementation is that all the internal nodes feature low impedance, minimizing the signal propagation delay from input to output. Thus, all the transistors are sized with small widths and lengths. The output transistors M_6 and M_8 drive the load capacitance with almost rail-to-rail output voltage, and are sized such that they can charge and discharge the output node fast enough to achieve the desired input-to-output delay.

The power consumption in the comparator implementation of Fig. 4.5 depends on the operating frequency of the PWM carrier signal and the load capacitance. The load capacitance (C_{load}) is typically the gate capacitor of a digital inverter in the range of 10-30 fF. Thus, the comparator power dissipation (P_{comp}) due to charging and discharging of the load capacitor and its quiescent current can be expressed as,

$$P_{comp} = P_Q + P_{SW} \cong V_{DD} \cdot I_{tail} + C_{load} \cdot V_{DD}^2 \cdot F_{SW}. \tag{4.15}$$

The goal of the comparator is to minimize its propagation delay by reducing the rise/fall time of its output. A rule-of-thumb is to target a rise/fall time of 0.5% of the switching period. For example, for a $F_{SW} = 500$ kHz with switching period of 2 μs, the desired maximum rise/fall time is 10 ns. To avoid being slew rate limited, the minimum I_{tail} needs to be higher than 5 μA, as expressed in Eq. (4.11).

If F_{SW} is increased to 2 MHz, then the minimum I_{tail} increases to 20 μA, leading to 4X times more power. Thus, increasing F_{SW} will directly increase the power consumption of the comparator.

Table 4.2 Design procedure for PWM comparator.

1.	Choose minimum I_{tail} to satisfy Eq. (4.11) and Eq. (4.12) for a given F_{SW} and $C_C = C_{load}$.
2.	Determine $(W/L)_1/(W/L)_3$ using Eq. (4.14) to maximize DC gain.
3.	Select M_1 transistor width, using minimum lengths to maximize Eq. (4.14)
4.	Select M_3 transistor width, using large lengths to maximize Eq. (4.14)
5.	If more DC gain is needed, $(W/L)_6/(W/L)_3$ can be increased.

Table 4.2 provides a concise design procedure for the comparator shown in Fig. 4.5. For example, $F_{SW} = 500$ kHz with switching period of 2 μs

requires a minimum $I_{tail} = 5\ \mu$A for a desired maximum rise/fall time of 10 ns in a $C_{load} = 10$ fF. For a DC gain of 60 dB or 1000 V/V, $(W/L)_1 = 1000 \cdot (W/L)_3$ is needed. Thus, the transistor sizes are selected as $(W/L)_1 = 18\ \mu\text{m}/0.18\ \mu\text{m}$ and $(W/L)_3 = 0.36\ \mu\text{m}/4\ \mu\text{m}$, having a DC gain of 60.91 dB or 1111 V/V. The triangle wave signal is typically generated from a relaxation oscillator as shown in Fig. 4.6, where the output triangle wave signal (V_{TRI}) is used as the input for two comparators that set the voltage limits of V_{TRI}. The output of the comparators are used to control a set-reset (SR) latch that generates the gate voltages for switches M_1 and M_2. The comparators used to set the voltage limits have to be fast and glitch free to avoid errors or peaking during the transitions. The SR-latch should have small metastability and small delay to avoid deviation from the desired F_{SW}.

The current sources should have high output resistance to avoid distortion of the carrier signal. The switch on-resistance have to be small enough to avoid voltage drops during each switching cycle.

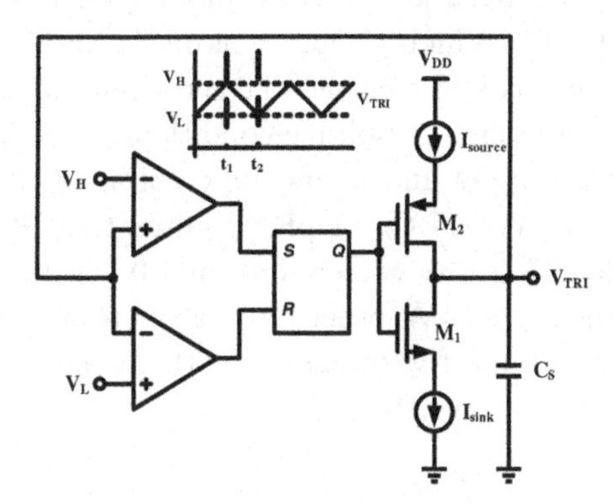

Fig. 4.6 Triangle wave generator circuit.

The operation is as follows: during the period t_1, M_2 is on and M_1 is off; thus, the capacitor C_S is charged with a constant current given by I_{source} to ramp up V_{TRI} until it exceeds the high voltage threshold V_H. Then, during the period t_2, M_1 is on and M_2 is off; thus, C_S is discharged with a constant current given by I_{sink} to ramp down V_{TRI} until it falls below the low voltage threshold V_L. This process is repeated each cycle and the period of the carrier wave signal is $T_{SW} = t_1 + t_2$. Thus, F_{SW} can be expressed as,

$$F_{SW,tri} = \frac{I_{sink}}{2C_S(V_H - V_L)} \tag{4.16}$$

where $I_{sink} = I_{source}$ is assumed. If the current sources are not equal, then the output triangle waveform will be distorted. The main design tradeoffs are between the area occupied by the capacitor and the current sources power consumption. A large C_S value allows low power consumption but occupies considerable area to implement it. A small C_S value occupies small area but consumes large power. Also, if I_{sink} is large, the drop across the on-resistor of the switches is more prominent, requiring large transistors to minimize the voltage drop.

The distortion of the output waveform results from the cropping of the peaks and valleys of V_{TRI}. If the output swing is rail-to-rail, then when the output signal is near a rail, the voltage drop across the current sources and the switches will limit the maximum output swing, distorting the output waveform. More details about how to analyze the distortion components of the resulting PMW signal are provided in Chapter 2.

The resulting $F_{SW,tri}$ is highly dependent on the variations of C_S. Thus, the capacitor choice needs to consider the sensitivity of the material used to implement it under different voltage conditions. The circuit is sensitive to variations in the supply and in its limit voltages. For this example, V_H=1.35 V, V_L=0.45 V, and C_S=1 pF, requires $I_{sink} \cong 1$ μA. V_{TRI} will have an amplitude of 0.9 V_{PP} centered around 0.9 V that is the common mode voltage of the system. Another alternative to minimize the effect of the variations is to use an active integrator with an amplifier at the expense of more power consumption.

4.4 Class-D output stage design

The class-D output stage is typically designed to minimize its dynamic power dissipation and its conduction power dissipation without degrading the propagation delay [74]. A typical implementation is shown in Fig. 4.7, where a non-overlapping signal generator is used to avoid excessive short-circuit current through the class-D output stage. The non-overlap delay (t_{ov}) is chosen as a tradeoff between efficiency and distortion [103]. Nevertheless, due to the large gate capacitance of the output stage transistors, a driver stage implemented as a tapered buffer is used to charge and discharge these large gate capacitances with minimum power-delay product [104].

For this example, nominal 1.8 V devices are used to implement the output power stage. Nonetheless, higher output power capabilities could be achieved using high-voltage devices such as thick-oxide, LDMOS, or DMOS transistors, depending on the desired application [56, 77].

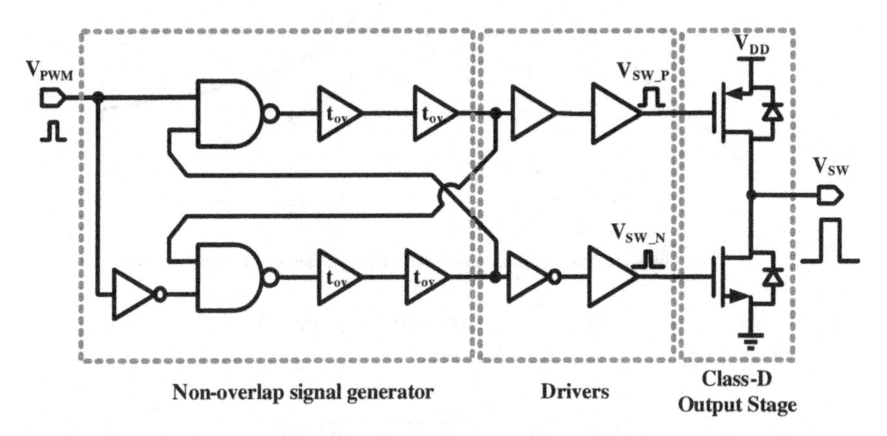

Fig. 4.7 Class-D output stage with auxiliary circuits.

The class-D output stage as shown in Fig. 4.7 is commonly known as a single-ended or half-bridge output stage. Its output transistors need to be carefully sized to avoid dynamic or conduction losses to dominate the overall efficiency performance. Figure 4.8 illustrates the design tradeoffs when choosing the width for the NMOS and PMOS transistors for a minimum length of 180 nm.

For example, for a typical EM speaker with an 8 Ω impedance and a desired efficiency at high output power of 97%, a $R_{dsON} = 0.2\ \Omega$ will be needed, as expressed in Eq. (1.12) and Eq. (2.2). Thus, from Fig. 4.8, the PMOS needs to have a width of 12 mm, and the NMOS needs a width of 3 mm to achieve the desired $R_{dsON} = 0.2\ \Omega$; this will result in gate capacitance for the PMOS of 16 pF and 4 pF for the NMOS from Fig. 4.8.

It can be noticed that further reducing the R_{dsON} will dramatically increase the gate capacitance, increasing the switching losses, and decreasing the power efficiency in the CDA at low output power. Since the transistors have large widths, their substrate area is large, resulting in a parasitic body-diode that plays a role in the power losses of the output stage, as expressed in Eq. (2.5). A sample design procedure for the half-bridge output stage is illustrated in Table 4.3. The maximum output power that the half-bridge output stage in Fig. 4.7 can provide to an EM speaker is a function of its

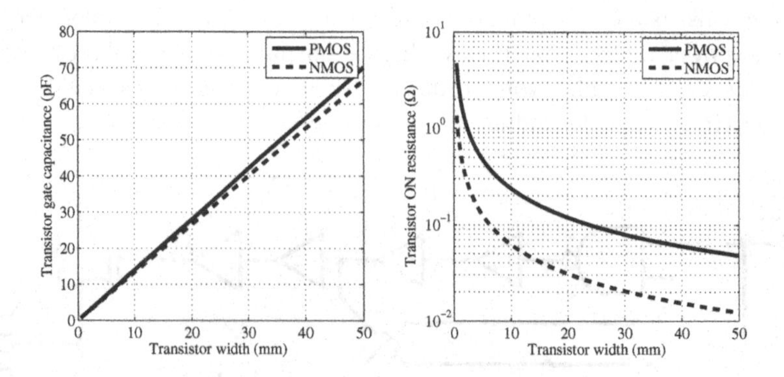

Fig. 4.8 Transistor gate capacitance and R_{dsON} versus transistor width.

supply voltage, load impedance, and finite R_{dsON}. The power rating can be expressed as,

$$P_{O,max} = V_{O,max} \cdot I_O = \frac{V_{O,max}^2}{R_{load} + R_{dsON}} = \frac{V_{O,PK}^2}{2 \cdot (R_{load} + R_{dsON})} \tag{4.17}$$

where $V_{O,max}$ is the maximum RMS voltage of the audio signal, and $V_{O,PK}$ is the maximum peak voltage of the audio signal. Note that the finite R_{dsON} will decrease the power rating of the amplifier. For this implementation example, the common mode level is 0.9 V for a supply voltage of 1.8 V; thus, the audio signal amplitude swings from 1.8 V to 0 V, resulting in a maximum peak voltage ($V_{O,PK}$) of 0.9 V. Then, $V_{O,max} = V_{O,PK}/\sqrt{2} =$ 0.636 V_{RMS}, resulting in a $P_{O,max}$ of 49.39 mW for an 8 Ω EM speaker with $R_{dsON} = 0.2$ Ω, or 96.42 mW for a 4 Ω EM speaker.

To increase the power rating of the CDA without increasing the supply

Table 4.3 Design procedure for half-bridge output stage.

Step	Procedure
1.	Determine maximum $I_{load} = V_{out}/\lvert Z_{load}\rvert$ for a given loudspeaker impedance and supply voltage.
2.	Choose desired efficiency region to optimize from Fig. 2.4.
3.	Plot the NMOS and PMOS transistor width versus $R_{ds,on}$ and gate capacitance, as in Fig. 4.8.
4a.	For region I, choose the NMOS and PMOS transistor width for the desired gate capacitance to minimize P_{SW}.
4b.	For region II, determine the NMOS and PMOS transistor width based on where the $R_{ds,on}$ and gate capacitance plot intersect each other.
4c.	For region III, choose NMOS and PMOS transistor width for the minimum $R_{ds,on} = R_{load}(1-\eta)/\eta$.

voltage, a full-bridge configuration as shown in Fig. 4.9 can be used as the class-D output stage, where the output audio signal is taken differentially as the voltage across the speaker load. This configuration is also known as a Bridge-Tied Load (BTL) or H-bridge output stage.

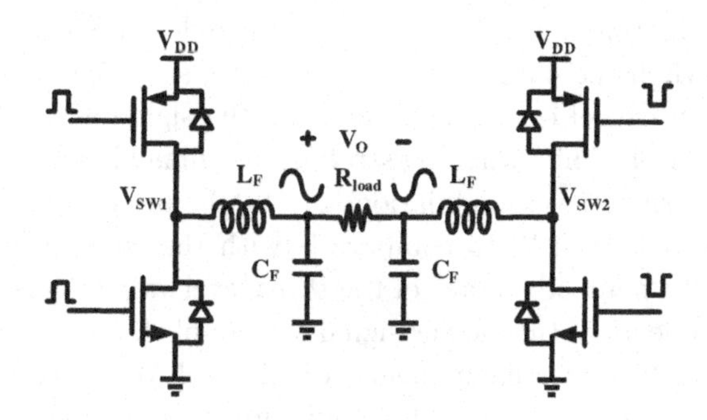

Fig. 4.9 Full-bridge class-D output stage.

Since each side or leg of the H-bridge processes the full audio signal amplitude but with 180° phase difference between each leg, the output voltage swing compared to the half-bridge output stage is doubled, increasing 4 times the output power. Also, the output signal even harmonics get ideally canceled by the subtraction operation, including the DC voltage. For the same example, with BTL $V_{O,max}$ is 1.272 V_{rms}, resulting in a $P_{O,max}$ of 202 mW for an 8 Ω EM speaker or 404.5 mW for a 4 Ω EM speaker.

To implement the H-bridge output stage, the active area used for the output switches and the cost of components in the output filter are doubled, since a complete half-bridge for each side is needed. Also, the CDA power consumption is doubled since the output signal is differential and needs to be translated to a single-ended configuration using an additional block in the feedback. A design procedure for the full-bridge output stage is presented in Table 4.4.

Another alternative is to use a differential CDA architecture, as observed in Fig. 4.10. Compared with the SE CDA architecture shown in Fig. 4.1, the differential CDA architecture doubles the amount of components needed to implement the system, increasing the area and power consumption of each block in the loop. Pseudo-differential or fully-differential amplifier configurations could be used in the compensator to process the two feedback paths [60, 105]. The main advantage is that the differential output voltage is processed differentially by the loop, enhancing greatly the audio performance.

One drawback in the H-bridge output stage is that, when the audio signal is not present, the switching signals ($V_{SW1,2}$) are still existing, and the differential switching signal ($V_{SW1} - V_{SW2}$) is changing from V_{DD} to $-V_{DD}$. Thus, some output current ripple is still present across the output filter, dissipating power. One alternative to decrease considerably this ripple is to switch the outputs $V_{SW1,2}$ in phase, such that when no signal is present both cancel each other, reducing the signal across the load, as shown in Fig. 4.11. This switching strategy is commonly known as filterless modulation or three level modulation [44, 78, 80], and its typical waveform is illustrated in Fig. 4.12 to compare it with the conventional H-bridge switching. The main advantage of the three level switching is that the frequency components of the carrier signal are doubled, as shown in Chapter 2, allowing smaller filter components, or, if the EM speaker has high inductance, to remove the external output filter at the expense of THD+N degradation and large EMI.

A more complex approach to cancel more harmonics in the switching output signal requires multiple power transistors at the output stage operating from two different supply voltages, as shown in Fig. 4.13 [45].

This approach achieves a multilevel switching signal which in turn increases the linearity of the quantization process by increasing the number of effective bits, as expressed in Eq. (3.8). Its switching output waveform is shown in Fig. 4.14 where 5 quantization levels can be observed.

The main drawback is the need for 8 switches, two floating supplies, and a complex switching strategy. The power consumption of this tech-

Table 4.4 Design procedure for full-bridge output stage.

Step	Procedure
1.	Determine maximum $I_{load} = 2V_{out}/\lvert Z_{load}\rvert$ for a given loudspeaker impedance and supply voltage.
2.	Choose desired efficiency region to optimize from Fig. 2.4.
3.	Plot the NMOS and PMOS transistor width versus $R_{ds,on}$ and gate capacitance, as in Fig. 4.8, for half-bridge of the output stage.
4a.	For region I, choose the NMOS and PMOS transistor width for the desired gate capacitance to minimize P_{SW}.
4b.	For region II, determine the NMOS and PMOS transistor width based on where the $R_{ds,on}$ and gate capacitance plot intersect each other.
4c.	For region III, choose NMOS and PMOS transistor width for the minimum $R_{ds,on} = R_{load}(1-\eta)/\eta$.
5.	Duplicate the designed transistors to create the full-bridge output stage.

nique dramatically limits the efficiency, making it not practical for mobile devices applications. Also, any timing error in the switching signals of any

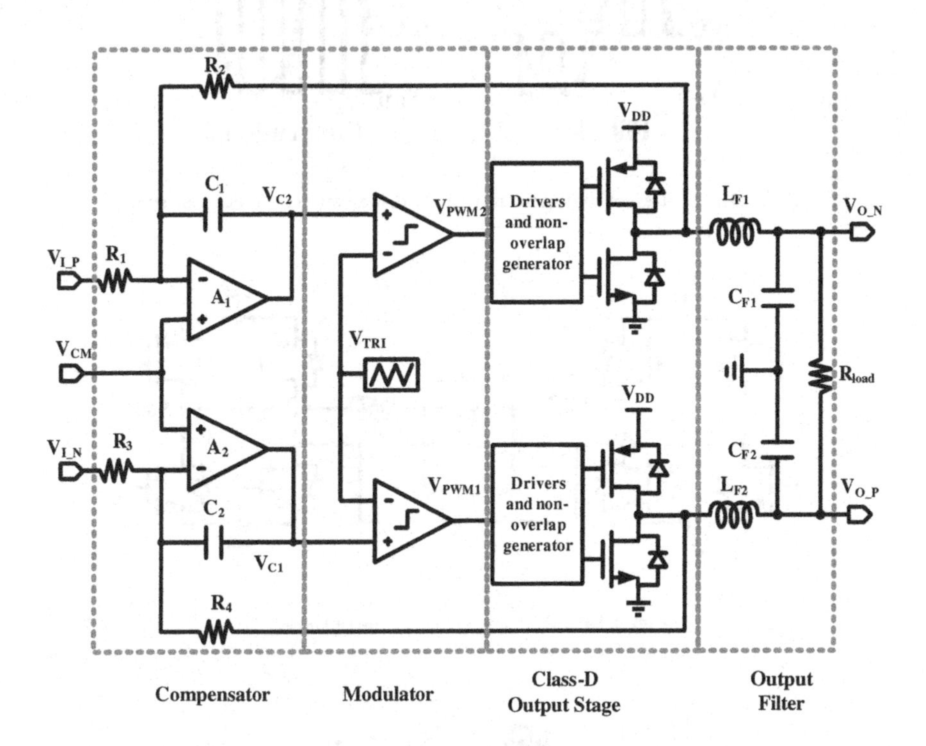

Fig. 4.10 First-order differential PWM CDA architecture.

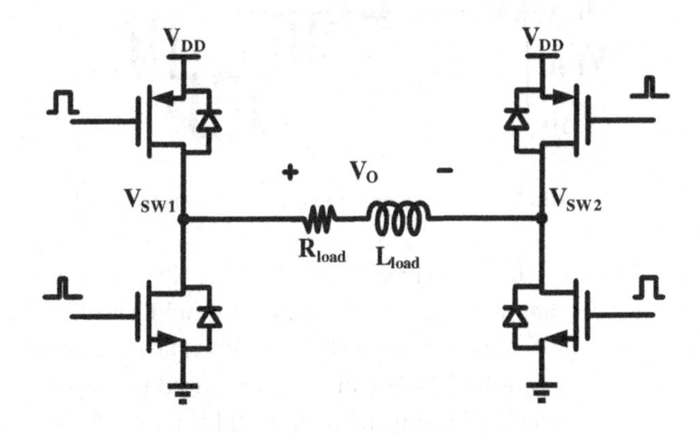

Fig. 4.11 Filterless class-D output stage.

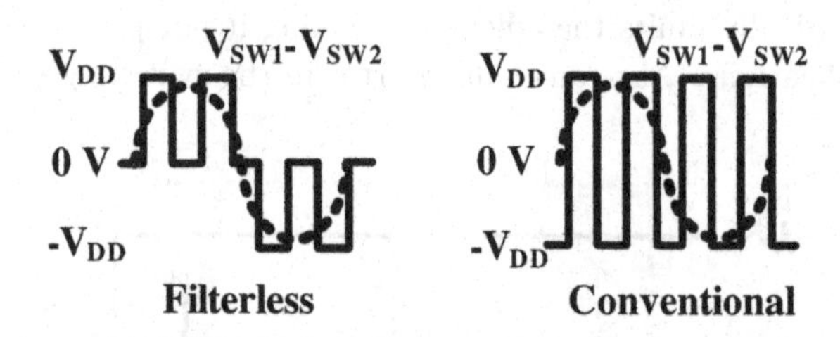

Fig. 4.12 Filterless and conventional H-bridge switching comparison.

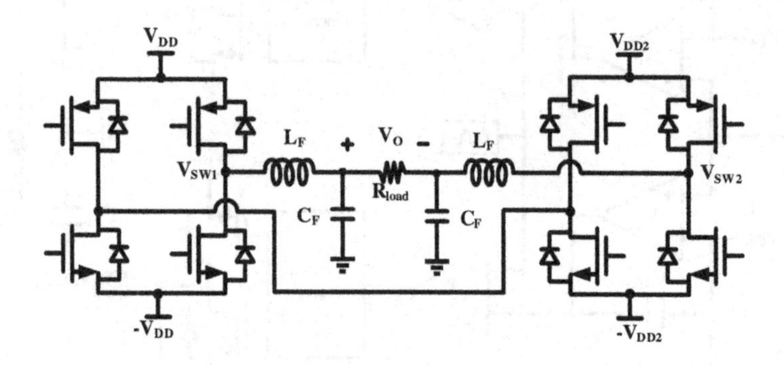

Fig. 4.13 Multilevel class-D output stage.

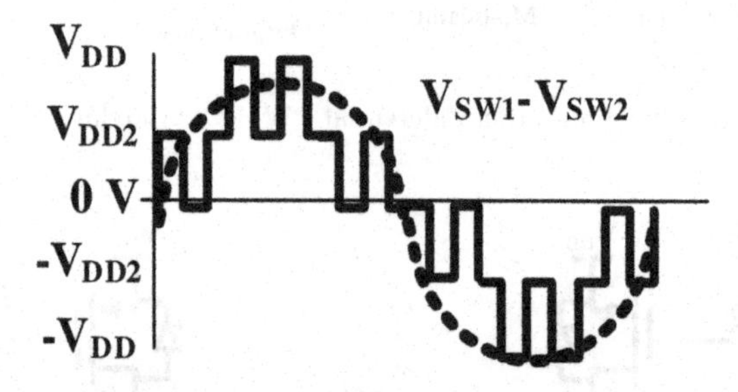

Fig. 4.14 Switching output waveform with 5 quantization levels.

transistor will reduce the effectiveness of the linearity enhancement. The effect of increasing the effective F_{SW} can also be achieved by using a single voltage supply but multiple phases in an interleaved output stage, as shown in Fig. 4.15 where 3 different phases separated 60° apart from each other are used to process the output signal. The main drawback of this output stage is that an extra inductor is needed for each additional clock

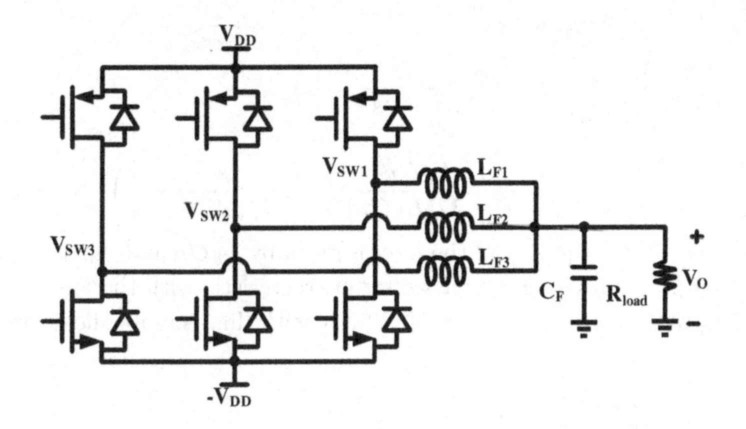

Fig. 4.15 Multiphase interleaved class-D output stage.

phase [106–108].

The multilevel and multiphase approach have also been used in DC-DC buck converters to reduce the output voltage and current ripple, and to lower the RMS current flowing into each inductor. However, since each additional clock phase or output level requires an extra inductor, the multiphase approach in Buck converters has been limited to 3 levels [109].

4.5 Output filter

The main function for the output filter is to recover the low frequency audio signal from the modulated output. An off-chip second-order low-pass filter is typically used. Since the output current flows through the output filter components, an inductor and a capacitor are typically used to avoid extra power dissipation. Figure 4.16 shows two typical configurations for the half-bridge output stage when dual supply or single supply are used.

The EM speaker must not be driven with a DC component to avoid damaging the transducer. Thus, the single supply configuration must have a large decoupling capacitor to remove the DC component. The transfer function for the LC filter with single supply and decoupling capacitor shown in Fig. 4.16 can be expressed as,

$$\frac{V_O(s)}{V_{SW}(s)}_{Single} = \frac{\omega_{LC}^2}{s^2+\omega_{LC}^2}\left(\frac{s}{s+\omega_z}\right)$$

$$= \frac{1/(L_F C_F)}{s^2+1/(L_F C_F)}\left(\frac{s}{s+1/(C_D R_{load})}\right) \tag{4.18}$$

where the cutoff frequency of the filter is given by $L_F C_F$ and the decoupling capacitor C_D creates a high pass filter characteristic with the load.

The transfer function for the LC filter with dual supply shown in Fig. 4.16 is expressed as,

$$\frac{V_O(s)}{V_{SW}(s)}_{Dual} = \frac{\omega_{LC}^2}{s^2+s\cdot 2\cdot\zeta\cdot\omega_{LC}+\omega_{LC}^2}$$

$$= \frac{1/(L_F\cdot C_F)}{s^2+s\cdot(R_{load}/L_F)+1/(L_F\cdot C_F)} \tag{4.19}$$

where it can be observed that the EM speaker equivalent resistance will determine the damping factor (ζ) of the LC filter.

The filter design presents several tradeoffs between the cut-off frequency, the F_{SW} of the system, the linearity of the components, and its power dissipation. The selection of the damping factor can affect the linearity of the audio signal if peaking is present in Eq. (4.19). Also, the inductor value selection will affect the amount of high frequency current ripple that will create power losses due to the non-idealities of the components.

For this example implementation, a dual supply filter was designed with a cutoff frequency of 22.5 kHz, and the filter components were chosen as $L_F = 50\ \mu$H, $C_F = 1\ \mu$F, and $R_{load} = 8\ \Omega$. This selection results in a Butterworth filter approximation which gives a flat magnitude and linear phase responses. For the single supply implementation, a blocking capacitor $C_D = 10\ \mu$F is needed to remove the DC component applied to the speaker, but the filter cutoff frequency remains the same as with the dual supply implementation.

4.6 Current and voltage sensor techniques

Some closed-loop CDA architectures need to monitor the output voltage and output current to determine the correct operation of the system [79, 81]. Also, when the current and voltage information is available, intelligent signal processing can be achieved to extract the real time impedance of the speaker, and calibrate the system to maximize the output power and efficiency [110]. Another useful function is the over-current protection of the output stage. The output voltage is easily monitored with the feedback loop since most CDA operate in voltage domain. Thus, a simple wire line connection is enough to sense the output voltage. The main challenge is to accurately monitor the output current. Several current sensing techniques have been explored in DC-DC power converters [111, 112]. Most of the techniques imply adding extra components in series or parallel to the output stage or the output filter to extract the output current information. A brief overview of the most used techniques will be discussed next.

The simplest method to measure the output current is to extract it from the inductor current using a sensing resistor (R_{sense}) in series with the inductor as shown Fig. 4.17, where the sensing voltage ($V_{sense} = I_O \cdot R_{sense}$) is frequency independent and proportional to I_O [79, 81].

The drawback of this technique is that R_{sense} is in the high current path, dissipating power. Also, its dynamic range is very limited since to sense small I_O, a large R_{sense} would be needed, but when I_O increases, the large R_{sense} would saturate the sensing amplifier output.

Another simple method is to use a current transformer or an inductor coupled to the inductor in the output filter, as illustrated in Fig. 4.18. This sensing technique dissipates very little power since the transformer is magnetically coupled to the inductor filter and attenuates the sensing current by its transformer gain determined by the N turns around the

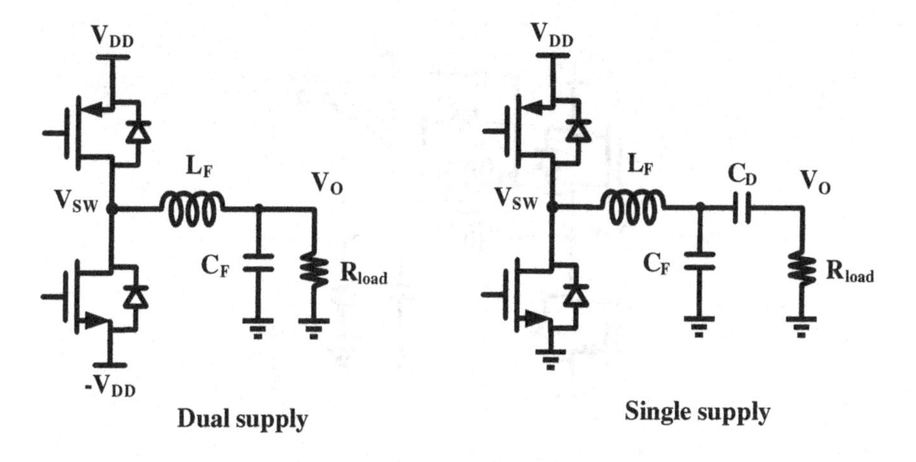

Fig. 4.16 Single-ended output filter configurations.

magnetic core; the sensing voltage is given by $V_{sense} = I_O/N \cdot R_{sense}$, where R_{sense} is used to convert the output current of the transformer to voltage.

The main drawback of this technique is that the current transformer is a bulky and expensive component that requires large PCB area to avoid EMI with other components. Also, its frequency response has to be high enough to process the carrier signal.

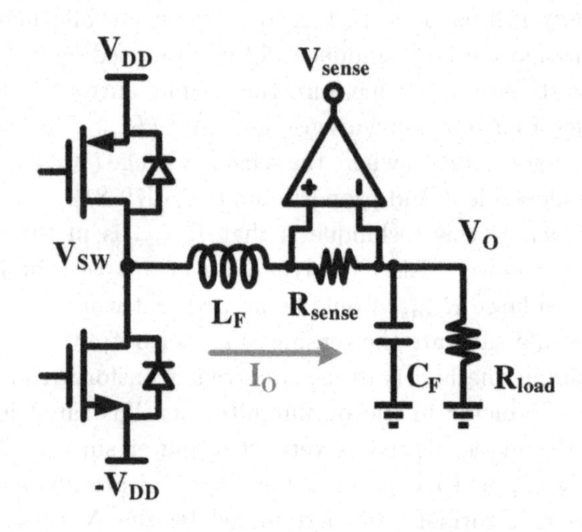

Fig. 4.17 Current sensing method using inductor series resistor.

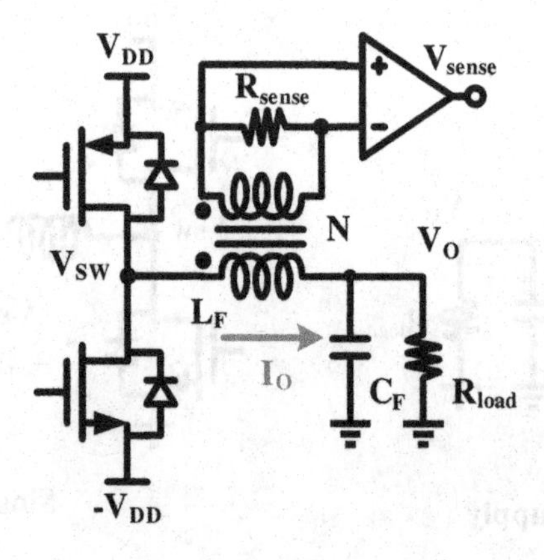

Fig. 4.18 Current sensing method using current transformer.

The sensing technique in Fig. 4.19 uses the parasitic inductor resistance (DCR) to sense the current across it. The DCR is not an explicit component and must be extracted from the inductor frequency response. In other words, the inductor and DCR have a high pass frequency response. Thus, a low pass frequency response is needed to flatten the frequency response, and extract the DCR magnitude which is proportional to the output current. The sensing voltage for the DCR sensing technique can be expressed as,

$$V_{sense} = \left(\frac{1 + sL_F/DCR}{1 + sR_{sense}C_{sense}}\right) \cdot DCR \cdot I_O. \tag{4.20}$$

The main drawback of this technique is that the inductor and DCR exact values are needed, and they depend on the manufacturer, that provides accuracies up to 20%. Thus, the technique is not accurate and highly sensitive to component variations.

All the previous current sensing techniques have the drawback of requiring external components. Moreover, the terminal capacitance, the lead inductance and resistance of each external component will affect the accuracy of the measurement. A current sensing technique that can be fully integrated on the same die as the output stage is desirable to achieve high accuracy in the sensing.

A fully integrated current sensor can be implemented using an on-chip sensing resistor in series with the NMOS transistor, as illustrated in Fig. 4.20. This resistor is typically implemented as a low value metal resistor. However, it only monitors the output current when the NMOS transistor is

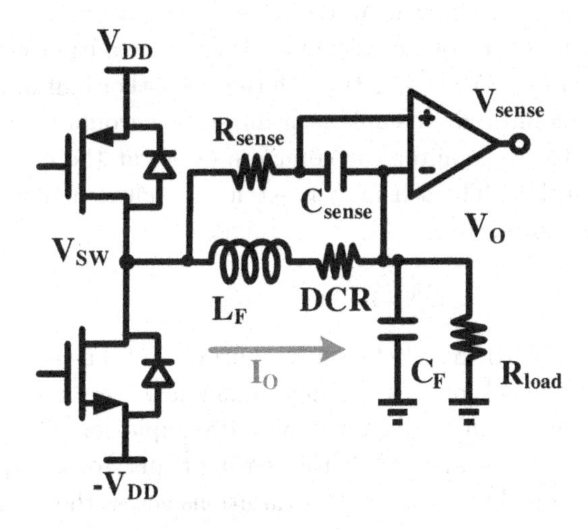

Fig. 4.19 Current sensing method using DCR sensing.

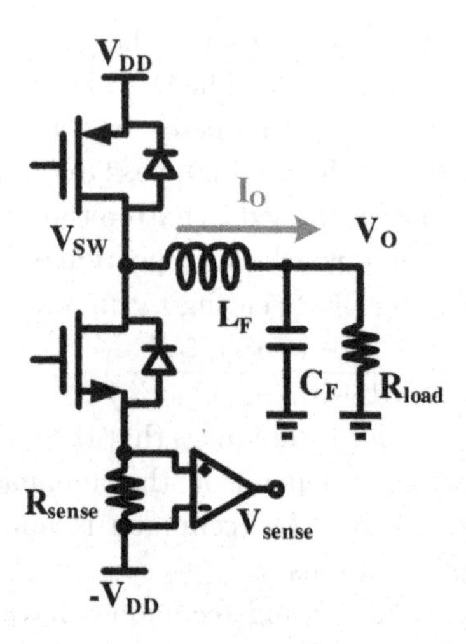

Fig. 4.20 Current sensing method using source sensing resistor.

active, and the sensing resistor dissipates power. This is a simple current sensing scheme that is typically preferred for over-current applications [113]. A more practical fully integrated sensing technique with good accuracy is shown in Fig. 4.21, where a sensing transistor K times smaller than the output transistor is used to extract the output current. The amplifier is used to force the same V_{DS} drop across both transistors. Since both transistor have the same V_{GS} and V_{DS}, the current flowing through R_{sense} is an attenuated copy of the current in the output transistor.

The main drawback of this technique is that the output current is only measured when the PMOS is active. During the other half of the cycle the current is not monitored. A similar sensing scheme would be needed for the NMOS transistor to monitor the complete cycle, at the expense of extra power consumption. The sensing voltage for the current mirror MOSFET sensor can be expressed as,

$$V_{sense} = \frac{I_O}{K} \cdot R_{sense} \tag{4.21}$$

where K is the width ratio between the output and sensing transistor.

Another drawback of this sensing technique is the matching between transistors. The output transistor is typically implemented as hundreds of small transistors in parallel, while the sensing transistor is implemented as a single transistor. Thus, the process variations across the chip affects each transistor differently, degrading the accuracy of the measurement. Also, the high frequency switching noise is passed to V_{sense} by the amplifier, requiring filtering to accurately sense small currents.

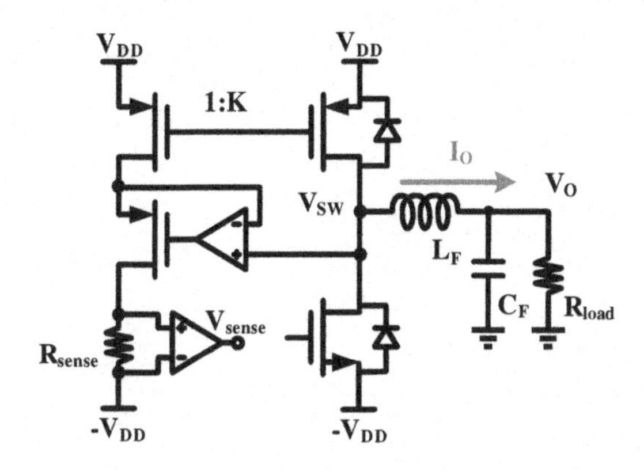

Fig. 4.21 Current sensing method using a current mirror MOSFET.

Table 4.5 summarizes the tradeoffs for each of the discussed current sensing techniques [111, 112]. Each technique is useful and has its applications. However, for CDA applications in mobile devices, the inductor series resistor, the transistor source resistor, and the transistor current mirror sensing techniques are the most suitable.

Table 4.5 Current sensing techniques comparison for CDA applications.

Sensing Technique	On-chip Integration	Accuracy	Power Dissipation	Complexity	Operational Bandwidth
Inductor series resistor [79]	No	H	H	L	H
Current Transformer [114]	No	H	L	L	L
Inductor DCR sensing [115]	No	M	L	H	L
Transistor source resistor [110]	Yes	M	H	L	M
Transistor current mirror [49, 113]	Yes	H	M	M	L

L=Low, M=Medium, H=High

Chapter 5

Power-Supply Noise Rejection Enhancement for Class-D Amplifiers

5.1 Power-supply noise in class-D amplifiers

Mobile devices often require the CDA output stage to be connected directly to the battery, providing the maximum amount of available power to the load [56, 72], as expressed in Eq. (4.17). In system-on-chip applications, the digital and radio frequency (RF) circuits often share the same battery as the analog circuits [57, 116]. Consequently, any noise on the battery power-supply plane is mixed together with the audio signal, degrading the amplifier's performance, as depicted in Fig. 5.1.

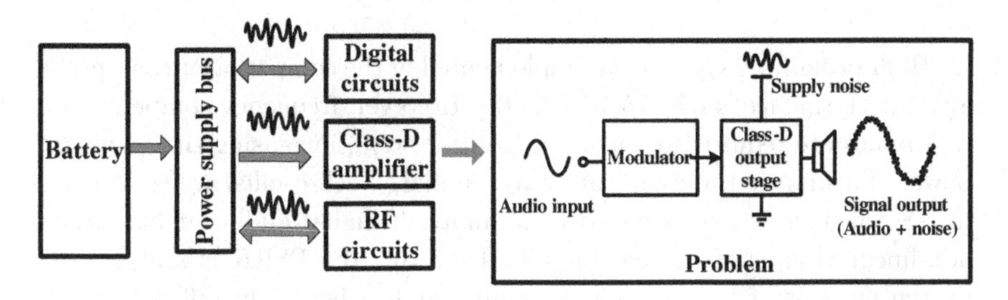

Fig. 5.1 Power supply noise problem in class-D amplifiers.

Open-loop CDA architectures are very sensitive to the power-supply noise since they directly couple the noise to the load each time the output stage switches. Hence, a good power supply rejection ratio (PSRR) is highly desirable for the CDA in battery-powered applications. One proposed solution to improve PSRR is to introduce a voltage regulator between the battery and the class-D amplifier to provide isolation from the power-supply noise [117]. However, this solution might reduce the available voltage delivered to the load, thereby limiting the audio amplifier's maximum output power. It also degrades the efficiency of the overall audio system due to the additional power dissipation in the voltage regulator.

Closed-loop CDA architectures conventionally enhance the PSRR by means of negative feedback, as discussed in Chapter 3. This feedback mechanism limits the noise coupled from the power-supply rail to the output. The resulting noise attenuation is proportional to the loop gain of the system. Thus, a common method to improve the PSRR is to attenuate the supply noise with a high-order compensator and large loop gain [60], as depicted in Fig. 5.2.

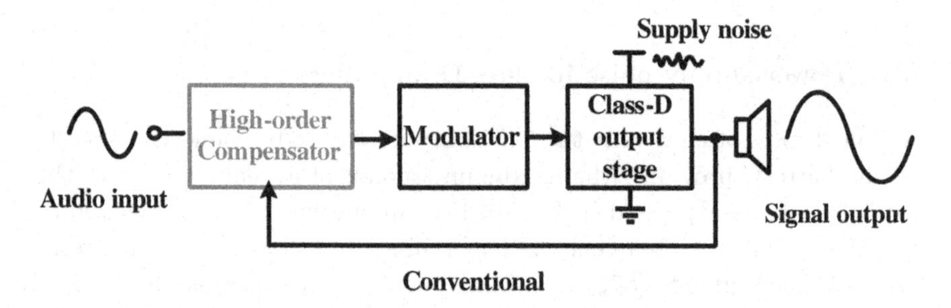

Fig. 5.2 Conceptual diagram of conventional solution to reduce power supply noise in class-D amplifiers.

High-order filters have been implemented in the compensator to improve audio performance [44,72,75,77,83,118]. However, frequency compensation techniques are required to ensure a stable system, increasing the quiescent power consumption and silicon area. Similarly, self-oscillating techniques in [78–83] have been suggested to accomplish higher audio quality, using non-linear compensation techniques. However, the PSRR is still limited by the amount of loop gain and/or mismatch achieved in differential architectures. Differential architectures with H-bridge output stages could provide good PSRR performance at the expense of larger silicon area and power consumption [44, 72, 75, 77, 83, 118]. In these architectures, matching between the differential paths limits the PSRR [60, 119]. Mismatch values as low as 0.01 % for passive components can be achieved with good layout, trimming, or calibration techniques [120]. However, this comes at the expense of increasing the silicon area and complexity. In [119] a self-adjusting voltage reference scheme was proposed to alleviate the matching requirements in bridge-tied load (BTL) differential architectures to achieve high PSRR. Nevertheless, the PSRR improvement is not constant across the audio bandwidth, limiting its benefits.

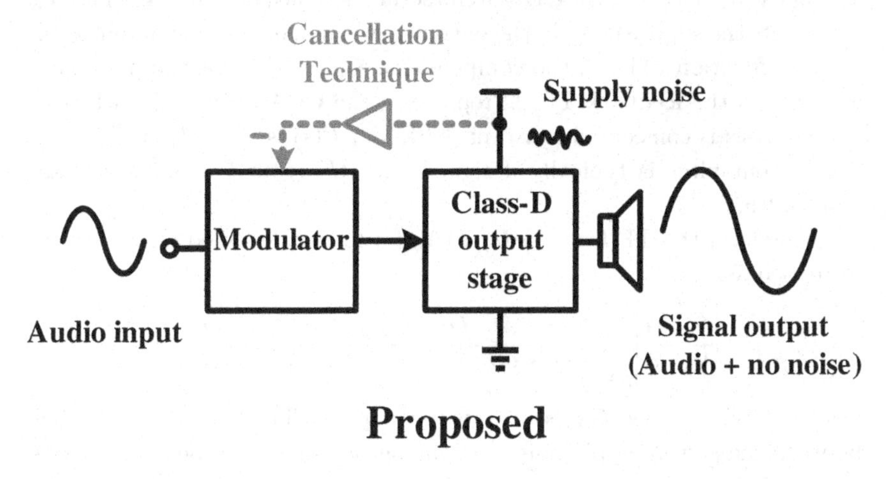

Fig. 5.3 Conceptual diagram of proposed solution to reduce power supply noise in class-D amplifiers.

This Chapter presents a design methodology to improve the PSRR in single-ended CDA architectures without increasing the compensator filter-order. This is accomplished by using a feed-forward power-supply noise cancellation (FFPSNC) technique [71] to suppress the supply-noise present in the system, as observed in Fig. 5.3. The technique improves the PSRR across the entire audio bandwidth independent of the compensator frequency characteristics. Also, the technique provides single-ended (SE) CDA architectures a PSRR performance comparable to differential architectures. Its small quiescent power and silicon area overhead, makes it an attractive alternative to enable high PSRR in the single-ended CDA architecture.

5.2 Power-supply noise modeling in class-D amplifiers

A review of the small signal linear model for the CDA is discussed to understand the system limitations. From these models, the power-supply noise transfer function for the CDA system is evaluated. The inputs of the system are assumed AC ground, and the only source of noise comes from the CDA output stage's power-supply. The SE CDA architecture and the differential BTL CDA architecture models are discussed to emphasize their tradeoffs.

5.2.1 *Single-ended load*

The linear model of the SE CDA architecture is shown in Fig. 5.4; where V_N represents the supply noise in the output power stage, G_C(s) symbolizes the transfer function (TF) of the compensator, G_M(s) denotes the modulator TF, β(s) is the feedback TF, D represents the CDA duty cycle (which in this analysis is considered constant [66]); and F(s) is the output filter TF. The output filter is typically designed with $|F(s)| \cong 1$ across the audio bandwidth.

Therefore, the TF from V_N(s) to V_O(s) for single-ended CDAs can be expressed as,

$$\left.\frac{V_O(s)}{V_N(s)}\right|_{SE} = \frac{D \cdot F(s)}{1 + G_C(s) \cdot G_M(s) \cdot \beta(s)} \cong \frac{D}{LG(s)} \tag{5.1}$$

where $LG(s) = G_C(s) \cdot G_M(s) \cdot \beta(s)$ is the loop gain TF. The ratio in decibels between the power-supply noise and the output signal can be expressed as,

$$\text{PSRR}_{SE,dB} = 20 \cdot \log\left(\left|\frac{V_N}{V_O}\right|\right)\Bigg|_{SE} \cong -20 \cdot \log(D) + 20 \cdot \log(|LG(s)|). \tag{5.2}$$

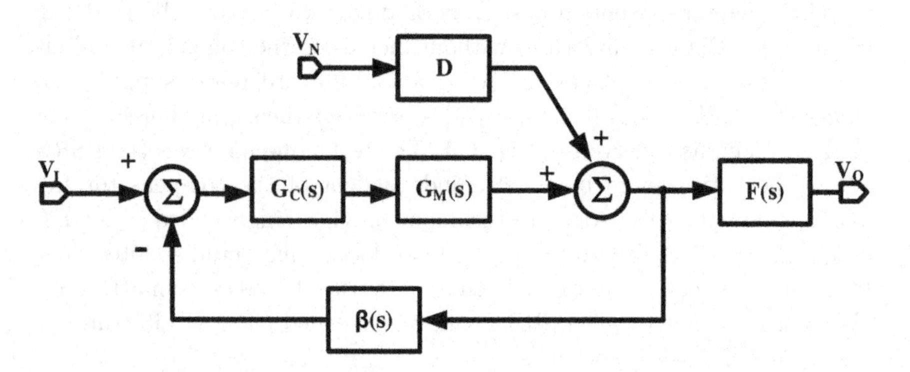

Fig. 5.4 Single-ended class-D audio amplifier linear model.

As can be seen in Eq. (5.2), large loop gain magnitude is required to have a high PSRR. As discussed in Chapter 3, the loop gain magnitude is typically increased by two methods. First, $G_C(s)$ could be enhanced by increasing the compensations filter-order. Nonetheless, stability for all the input signal magnitude range is more difficult to achieve in high-order filters, and would require large silicon area and extra quiescent power consumption [44,72,75,77,83].

Second, $G_M(s)$ could be increased to enhance PSRR. For example, a constant frequency PWM modulator is modeled as a linear constant magnitude, depending only on its input [66]; as expressed by,

$$G_M = \frac{V_{supply}}{V_{TRI}} \tag{5.3}$$

where V_{supply} denotes the modulator's output square wave voltage amplitude and V_{TRI} is the triangular-wave carrier's peak-to-peak voltage amplitude. To increase G_M in a battery-powered device with fixed V_{supply}, V_{TRI} needs to be reduced. This requires a more stringent design on the comparator to detect smaller voltages, and as a result, the PSRR improvement is limited. Other modulation schemes like self-oscillating modulation (SOM) or sigma-delta modulation (SDM) will have different $G_M(s)$, providing a different gain in the loop.

5.2.2 *Bridge-tied load*

The linear model for the BTL differential CDA architecture is illustrated in Fig. 5.5. It can be observed that each differential path receives the same noise contribution from the power-supply, assuming ideal matched conditions between the two paths. Therefore, we can express the transfer function for the power-supply noise of the BTL CDA to its differential output $V_O(s) = V_{O,P}(s) - V_{O,N}(s)$ as,

$$\left.\frac{V_O(s)}{V_N(s)}\right|_{BTL} = \frac{D \cdot (LG_2(s) - LG_1(s)) \cdot F(s)}{1 + LG_1(s) + LG_2(s) + (LG_2(s) \cdot LG_1(s))}$$

$$\cong D\left(\frac{LG_2(s) - LG_1(s)}{LG_2(s) \cdot LG_1(s)}\right) \tag{5.4}$$

where $LG_i(s) = G_{Ci}(s) \cdot G_{Mi}(s) \cdot \beta_i(s)$, for $i = 1, 2$.

It can be observed in the numerator of Eq. (5.4) that the power-supply noise contribution depends on the difference between $LG_1(s)$ and $LG_2(s)$. Hence, assuming $LG(s) = LG_2(s) = (1 \pm \delta) \cdot LG_1(s)$ and $0 < \delta < 1$, to take into account the deviation δ between the two differential paths, the transfer function in Eq. (5.4) becomes,

$$\left.\frac{V_O(s)}{V_N(s)}\right|_{BTL} \cong D\left(\frac{(1 \pm \delta)LG_1(s) - LG_1(s)}{LG(s) \cdot LG_1(s)}\right) \cong D\left(\frac{|\delta| \cdot LG_1(s)}{LG(s) \cdot LG_1(s)}\right)$$

$$\cong D\left(\frac{|\delta|}{LG(s)}\right). \tag{5.5}$$

The PSRR transfer function in decibels for the BTL CDA yields,

$$PSRR_{BTL,dB} = 20 \cdot \log\left(\left|\frac{V_N}{V_O}\right|\right)\Big|_{BTL}$$
$$\cong -20 \cdot \log(D) + 20 \cdot \log(|LG|) - 20 \cdot \log(|\delta|) \quad (5.6)$$

In the ideal scenario where both paths are perfectly matched, this architecture provides infinite PSRR since the deviation (δ) would be zero. Nonetheless, mismatch geometries on passive components and active devices, together with amplifier errors (due to amplifier finite loop-gain and bandwidth), will limit the PSRR performance. In other words, the deviation δ is a frequency dependent variable since it needs to match the frequency responses of both differential paths.

For example, a second-order BTL CDA design with a deviation δ = 10% between both loops, $|LG(s)|$ = 70 dB at 217 Hz, and D=0.5 would have a PSRR of 96 dB at low frequencies according to Eq. (5.6). The deviation δ is typically the limiting factor for the PSRR in differential CDA architectures, and it depends on the matching of two passive feedback networks (e.g. for a first-order compensator shown in Fig. 4.10, the error δ would be limited by the mismatch between two resistors and two capacitors [60]). Also, if the modulator is implemented with a pseudo-differential arrangement of comparators, then the delay, offset, and gain mismatch between comparators will increase the deviation δ in the loop.

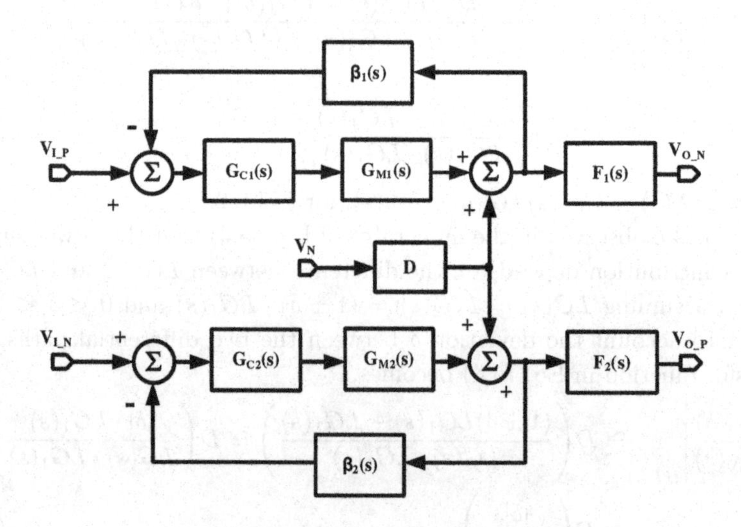

Fig. 5.5 Bridge-Tied Load differential CDA linear model.

5.3 Feed-forward power-supply noise cancellation technique

The proposed feed-forward power-supply noise cancellation (FFPSNC) technique linear model is illustrated in Fig. 5.6 for the CDA architecture. As can be observed, an additional feed-forward path $G_{FF}(s)$ is introduced in the system to inject the power-supply noise at the input of the modulator block. The feed-forward path's purpose is to replicate the power-supply noise with the correct gain and polarity and inject it into the system to cancel out the supply noise going through the feedback loop before it reaches the modulator block. This is because the modulator block performs a non-linear operation that results in intermodulation and harmonic distortion.

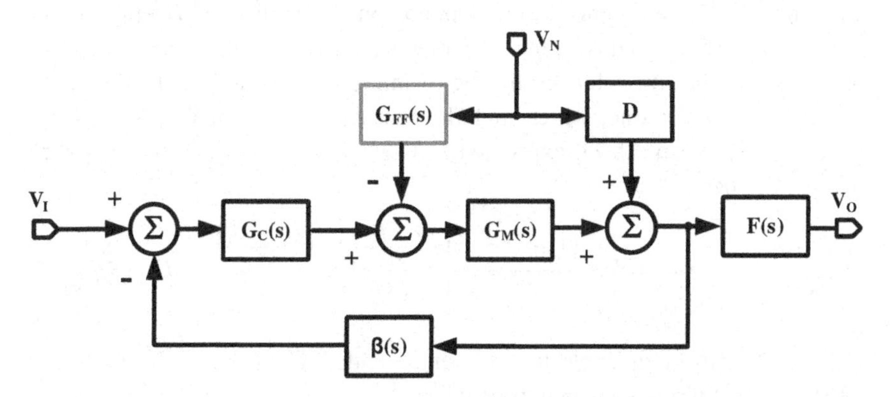

Fig. 5.6 Class-D amplifier linear model with proposed FFPSNC technique.

The transfer function from $V_N(s)$ to $V_O(s)$ for the CDA (including the proposed FFPSNC technique) is given by,

$$\left.\frac{V_O(s)}{V_N(s)}\right|_{FFPSNC} = \frac{D \cdot \left(1 - G_{FF}(s) \cdot \left(\frac{G_M(s)}{D}\right)\right) \cdot F(s)}{1 + G_C(s) \cdot G_M(s) \cdot \beta(s)}. \tag{5.7}$$

It is worth noting that if we remove the $G_{FF}(s)$ path, the transfer function in Eq. (5.7) reduces to Eq. (5.1). By observing the numerator of Eq. (5.7), we can conclude that the power-supply noise present at the output would be completely canceled by selecting $G_{FF}(s) = D/G_M(s)$. This cancellation can be achieved independently of $G_C(s)$. Another alternative to implement the feed-forward path $G_{FF}(s)$ is to apply it before the compensator block $G_C(s)$. However, the required $G_{FF}(s)$ would contain the reciprocal of $G_C(s)$, which is a frequency dependent block, and would require matching more components, making this choice less feasible. Therefore, the proposed path shown in Fig. 5.6 is the best tradeoff choice between additional hardware overhead and design complexity. Note that the FFPSNC technique could be applied to BTL architectures but it would require two feedforward paths applied to each differential path, introducing another mismatch element. The PSRR would be limited by the differential path mismatch, the mismatch between the feed-forward cancellation paths, and the $G_{FF}(s)$ implementation mismatch, making the technique not very feasible for BTL implementations.

5.3.1 *System analysis*

The amount of noise cancellation depends on the ability of $G_{FF}(s)$ to precisely replicate $D/G_M(s)$. Thus, any deviation from it will limit the amount of noise cancellation achieved by the technique. Considering the deviation α from the ideal value, which includes variations in D or $G_M(s)$, the actual $G_{FF}(s)$ in Eq. (5.7), expressed as $G_{FF}(s) = (1 \pm \alpha) \cdot D/G_M(s)$ with $0 < \alpha < 1$, becomes,

$$\left.\frac{V_O(s)}{V_N(s)}\right|_{FFPSNC} = D \cdot \frac{(1 - (1 \pm \alpha) \cdot (\frac{D}{G_M(s)}) \cdot (\frac{G_M(s)}{D}))}{1 + LG(s)} \cong D \cdot \frac{|\alpha|}{LG(s)}. \tag{5.8}$$

The PSRR in decibels for the single-ended CDA with the proposed FFPSNC technique can be expressed as,

$$\begin{aligned} \text{PSRR}_{FFPSNC,dB} &= 20 \cdot \log\left(\left|\frac{LG(s)}{D \cdot \alpha}\right|\right) \\ &\cong -20 \cdot \log(D) + 20 \cdot \log(|LG|) - 20 \cdot \log(|\alpha|). \end{aligned} \tag{5.9}$$

The PSRR improvement for the proposed technique is $-20 \cdot \log(|\alpha|)$ compared with Eq. (5.2). The FFPSNC technique provides the benefit of additional PSRR limited by the amount of mismatch between two paths, as in BTL architectures. However, to minimize the error δ in Eq. (5.6), a precise match of two feedback networks must be obtained to achieve high PSRR, and this matching would consume a large silicon area [60,105]. In the proposed FFPSNC technique, the deviation α will depend on the implementation of $G_{FF}(s)$.

The key parameter for the implementation of $G_{FF}(s)$ is a good extraction of the linear gain $G_M(s)$ since it varies across different modulation schemes. However, the modulation process is a non-linear operation that depends on its input amplitude, and it requires a quasi-linearization to be able to extract the equivalent gain $G_M(s)$. To accomplish this, the describing function (DF) methodology is applied because it provides an approximate procedure for analyzing certain non-linear blocks in control systems such as the modulation of switching circuits [86, 121]. A general representation for the DF as a complex gain for a sinusoidal input of amplitude V_i and frequency ω can be expressed as,

$$\begin{aligned} G_M(V_i,\omega) &= G_p(V_i,\omega) + jG_q(V_i,\omega) \\ &= M_G(V_i,\omega)e^{j\phi_G(V_i,\omega)} \end{aligned} \tag{5.10}$$

where the terms $G_p(V_i,\omega)$ and $G_q(V_i,\omega)$ are the in-phase and quadrature gains of the non-linearity. The magnitude and phase representation can be expressed as,

$$\begin{aligned} M_G(V_i,\omega) &= \sqrt{G_p^2(V_i,\omega) + G_q^2(V_i,\omega)}, \\ \phi_G(V_i,\omega) &= \tan^{-1}\left(\frac{G_q(V_i,\omega)}{G_p(V_i,\omega)}\right). \end{aligned} \tag{5.11}$$

The modulation schemes used in the CDA are typically implemented with a relay, relay with hysteresis, or odd quantizer non-linearities. For the PWM, the modulation is implemented with a comparator with sharp transition that can be approximated to a relay non-linearity [86] with a DF given by,

$$G_M(V_i,\omega)_{PWM} = \frac{V_{DD}}{V_i} \cong \frac{V_{DD}}{V_{TRI}} \tag{5.12}$$

where V_{DD} is the supply of the comparator and V_{TRI} is the peak amplitude of the triangular carrier waveform. For the close loop PWM CDA architecture, the loop forces the input of the comparator to be within the carrier peak amplitude. Thus, the quasi-linearized gain in Eq. (5.12) using V_{TRI} as the peak amplitude at the input of PWM modulator corresponds to the expression in Eq. (5.3), verifying the analysis.

Also, it is worth noticing that Eq. (5.12) is frequency independent, which allows a simpler $G_{FF}(s)$ implementation expressed as,

$$G_{FF}(s)_{PWM} = \frac{D}{G_{M,PWM}} \cong \frac{D \cdot V_{TRI}}{V_{DD}}. \tag{5.13}$$

For the SDM, the modulation is implemented with a quantizer that can be approximated to an uniform quantizer non-linearity with a DF given by,

$$G_M(V_i,\omega)_{SDM} = \frac{V_{DD}}{V_i}\sum_{m=1}^{n}\sqrt{1-\left(\frac{2m-1}{2}\frac{q}{V_i}\right)^2} \tag{5.14}$$

where q is the quantization step as expressed in Eq. (3.6), and n is the number of quantization output levels in the quantizer. It can be observed that as n increases, the G_M approaches a linear gain. This is expected since a multilevel quantizer provide less quantization error and a more linear operation as expressed in Eq. (3.7), at the expense of increased power consumption as discussed in Chapter 4.

The gain in Eq. (5.14) is frequency independent and will approach unity for $V_i \gg q$, assuming $V_i = V_{DD}$ for full dynamic range. However, since the quantizer in a SDM architecture is typically preceded by a sample and hold circuit, the modulator gain $G_{M,SDM}(s)$ contains the delay introduced by this block. The sample and hold operation can be approximated to a zero-order hold (ZOH) model with transfer function expressed as,

$$G_{ZOH}(s) = \frac{1 - e^{-sT_{SW}}}{sT_{SW}} \tag{5.15}$$

where $T_{SW} = 1/F_{SW}$. The implementation of $G_{FF}(s)$ for SDM needs to consider both gains and it would be given by,

$$G_{FF}(s)_{SDM} = \frac{D}{G_{M,SDM}} \left(\frac{1}{G_{ZOH}(s)} \right). \tag{5.16}$$

To evaluate the effect of the ZOH, a bode plot for the ZOH with two different sampling frequencies is shown in Fig. 5.7. It can be observed that if the sampling frequency F_{SW} is much higher than the desired signal bandwidth, the magnitude of Eq. (5.15) is almost unity. However, the phase shift is dramatically different for a $F_{SW} = 200\ kHz$ with a phase shift of -20° at 20 kHz bandwidth, while for a $F_{SW} = 2\ MHz$, the phase shift is -1.8°. Thus, for $F_{SW} > 2\ MHz$ the ZOH transfer function can be obviated.

For the SOM, the modulation is implemented with a hysteretic comparator that can be approximated to a relay with hysteresis non-linearity with a DF defined in Eq. (3.14). Its modulator gain, magnitude, and phase can be expressed as [86],

$$\begin{aligned} G_M(V_i, \omega)_{SOM} &= \frac{V_{DD}}{V_h} e^{j \tan^{-1}\left(\frac{V_h/V_i}{\sqrt{1 - (V_h/V_i)}} \right)}, \\ M_G(V_i, \omega)_{SOM} &= \frac{V_{DD}}{V_h}, \\ \phi_G(V_i, \omega)_{SOM} &= \tan^{-1}\left(\frac{V_h/V_i}{\sqrt{1 - (V_h/V_i)}} \right) \cong \sin^{-1}\left(\frac{V_h}{V_i} \right) \end{aligned} \tag{5.17}$$

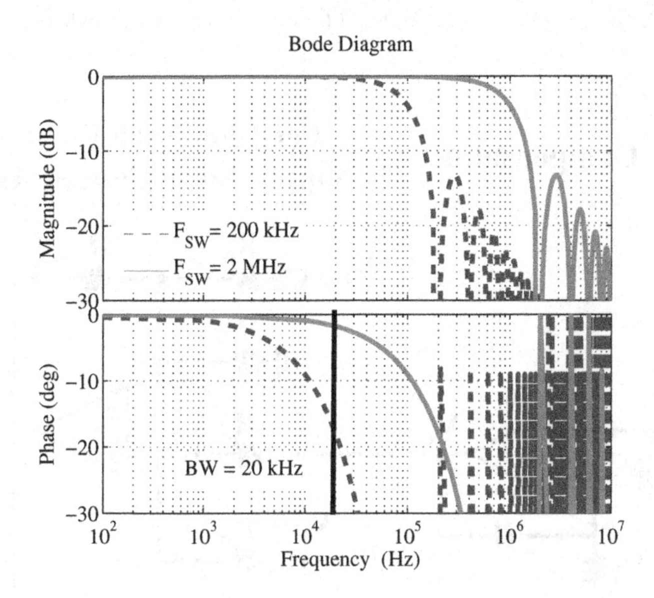

Fig. 5.7 Magnitude and phase frequency response of zero-order hold.

where V_h is the hysteresis window. As in the PWM case, the close loop will force the input signal of the hysteretic comparator to be within the hysteresis window. Thus, the $M_G(V_i,\omega)_{SOM}$ is constant and frequency independent, but its phase response is a function of the input amplitude and the hysteresis window.

Contrary to the PWM or SDM, the SOM is highly dependent on the modulator input voltage and would present a input-dependent delay. Thus, the $G_{FF}(s)$ implementation has to replicate the reciprocal of this phase variation, and it can be expressed as,

$$G_{FF,SOM}(s) = \frac{D}{G_{M.SOM}(s)} \cong \frac{D \cdot V_h}{V_{DD} \cdot \left(\sqrt{1-\left(\frac{V_h}{V_i}\right)^2} + \frac{V_h}{V_i}s\right)}. \tag{5.18}$$

5.3.2 *Circuit implementation*

From all the discussed modulation schemes, the PWM provides the simpler implementation for $G_{FF}(s)$. Thus, to demonstrate the effectiveness of the proposed scheme, a first-order SE PWM CDA with the FFPSNC technique is implemented. Figure 5.8 shows the schematic circuit of the implemented

CDA with the proposed technique. The design for each block is addressed next.

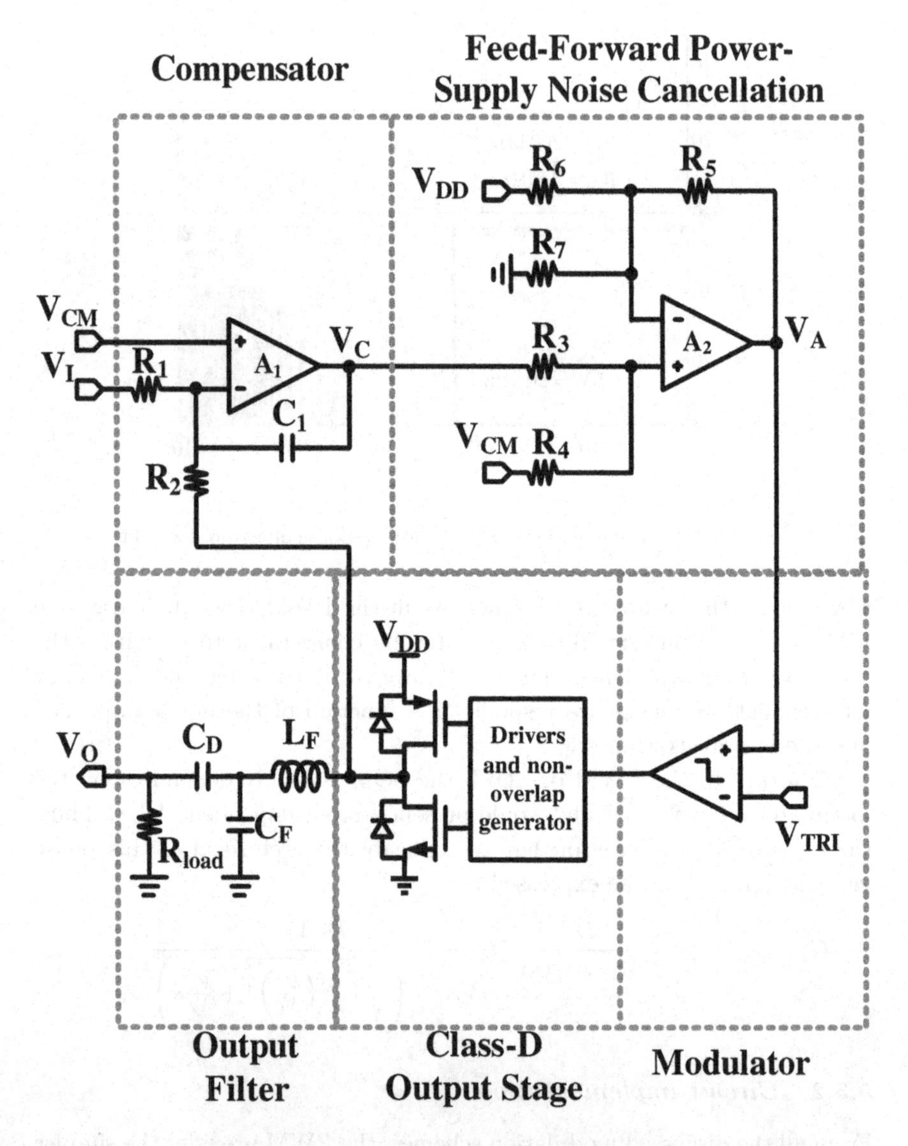

Fig. 5.8 Proposed CDA implementation with FFPSNC technique.

The compensator was implemented as a first-order continuous-time integrator with crossover frequency $f_{int} = 1/(2\pi \cdot R_1 \cdot C_1)$. Its value selection depends on several tradeoffs between the passive's area, amplifier's power, linearity, and noise, as discussed in previous chapters. To avoid significant deviations in the compensators performance, the amplifier's gain-bandwidth product (GBW) has to be higher than f_{int}, as expressed in Eq. (4.1).

As discussed in Chapter 3, a large value for f_{int} would provide a higher bandwidth for the CDA loop. A large bandwidth would result in high linearity, high PSRR at higher frequencies, and smaller passive component values. However, the high frequency amplifier's noise would also be amplified, and the amplifier's power would need to be increased to avoid deviations in f_{int} due the finite GBW [101]. The f_{int} = 15.5 kHz was chosen as a compromise between these tradeoffs.

The integrator component values are C_1 = 32 pF and $R_1 = R_2$ = 320 $k\Omega$. The input resistor values were chosen considering the tradeoff between the resistor's thermal noise contribution and its matching requirements. The amplifier A_1 is implemented as a two-stage Miller-compensated amplifier, designed as discussed on Chapter 4, where the input transistors used lengths of 2 μm to minimize their flicker noise contribution in the audio bandwidth. The amplifier A_1 achieves a DC open-loop gain of 76 dB with a phase margin of 61°, and GBW of 10 MHz. The amplifier consumes a quiescent current of 28 μA from a 1.8 V supply (V_{DD}).

Pulse width modulation (PWM) was chosen for this implementation since its quasi-linear modulation gain, $G_M(s)$, can be approximated as a constant in the audio bandwidth if the modulation frequency is constant and at least two times higher than the audio bandwidth [66]. The G_M magnitude represents the linear gain of the combination of the modulator and output power stage. Note that the modulator's noise contribution from the supply is already represented as the noise signal at the output of the linear model.

The PWM modulator was implemented using an open-loop comparator with large open-loop gain. The comparator was designed as a push-pull amplifier, as show in Fig. 4.5. The comparator consumes 20 μA of quiescent current from a V_{DD} of 1.8 V. An external 500 kHz triangle-wave carrier signal with peak-to-peak amplitude of 0.9 V was used to have a modulator gain $G_M \approx 2$. This is to have external control on the amplitude of the triangular waveform for manual calibration of G_M.

The output power stage was designed to minimize dynamic power dissipation without degrading the propagation delay [74], as discussed in

Chapter 4. The implementation was shown in Fig. 4.7.

For demonstration of the FFPSNC technique, nominal 1.8 V devices were used to implement the output stage. Nonetheless, higher output power capabilities could be expanded using high voltage devices such as thick oxide, LDMOS, or DMOS transistors depending on the desired application [56, 77].

The main function for the output filter is to recover the low frequency audio signal from the modulated output. An off-chip second-order low-pass filter is typically used, as explained in Chapter 4. The filter components were implemented as $L_F = 50\ \mu$H, $C_F = 1\ \mu$F, and $R_{load} = 8\ \Omega$. A blocking capacitor $C_D = 10\ \mu$F was selected to remove the DC component applied to the speaker.

For the implemented PWM scheme, $G_{FF}(s)$ is assumed constant since the average magnitude of D and G_M, with no input signal, is assumed constant [66]. Then, G_{FF} is implemented based on a resistor's ratio. To minimize silicon area and quiescent power consumption, the FFPSNC technique was implemented using an additional amplifier A_2 in a balanced adder configuration as shown in Fig. 5.8. Assuming A_2 is an ideal amplifier and that the supply noise voltage V_N comes from V_{DD}, the output of the amplifier can be expressed as,

$$
\begin{aligned}
V_A &= V_C \cdot \left[\left(\frac{R_4}{R_4 + R_3}\right) \cdot \left(1 + \frac{R_5}{R_6//R_7}\right)\right] - V_N \cdot \left(\frac{R_5}{R_6}\right). \\
&= \frac{V_C}{1 + \frac{R_3}{R_4}} \left(1 + \frac{R_5}{R_7} + \frac{R_5}{R_6}\right) - V_N \cdot \left(\frac{R_5}{R_6}\right). \qquad (5.19)
\end{aligned}
$$

This arrangement allows the use of only one amplifier to provide two functions: 1) provide a feed-forward path to add V_N to the system, and 2) scale V_N with the proper gain and polarity. Resistor ratio R_5/R_6 implements G_{FF}, and their values were chosen for an average $D = 0.5$ such that $G_{FF} = D/G_M = 0.5/2 = 0.25$. The values of resistors $R_3 - R_4$ were chosen to provide a unity gain path from V_C to V_A, to avoid altering the feedback loop characteristics. Resistor R_7 value was chosen equal to R_6 to set the DC value of the negative input of A_2, at the system's common-mode voltage of 0.9 V.

For the proposed implementation, the deviation α in Eq. (5.9) can be minimized without a large silicon area requirement. Also, since the PWM has a $G_M(s)$ constant across the audio bandwidth, then the FFPSNC technique would be effective across the entire audio bandwidth.

5.4 Design tradeoffs and methodology

The proposed FFPSNC implementation for $G_{FF,i} = -(R_5/R_6)$ in Eq. (5.19) presents important design choices and tradeoffs. To evaluate the effects of mismatch in the implementation of R_5/R_6 resistors and the gain error due to finite loop gain in the amplifier A_2, the deviation α in Eq. (5.9) can be broken down in two error components.

First, the resistor mismatch in R_5/R_6 is expressed as $(R_5/R_6)\cdot(1\pm\alpha_1)$, for $0 < \alpha_1 < 1$. Second, the amplifier's A_2 DC gain error due to finite loop gain is expressed as $\varepsilon = 1/(A_2(s)\cdot\beta_{FF})$, for $(A_2(s)\cdot\beta_{FF}) \gg 1$, where $A_2(s)$ is the amplifier's open-loop gain and β_{FF} is the feedback gain [101].

The gain error ε in A_2 is further divided since the feedback gain is expressed as,

$$\beta_{FF} = \frac{R_6//R_7}{R_5 + (R_6//R_7)}, \tag{5.20}$$

that also contains the mismatch component α_1 between R_5/R_6. Considering both error components, the implemented feed-forward gain $G_{FF,a}$ is,

$$\begin{aligned} G_{FF,a}(s) &= \frac{G_{FF,i}(s)}{1 + \frac{1}{A_2(s)\cdot\beta_{FF}}} \\ &\cong \frac{-\left(\frac{R_5}{R_6}\right)(1\pm\alpha_1)}{1 + \frac{1}{A_2(s)}\left[1 + \left(\frac{R_5}{R_6}\right)(1\pm\alpha_1)\left(1+\frac{R_6}{R_7}\right)\right]}. \end{aligned} \tag{5.21}$$

Assuming $|A_2(s)\cdot\beta_{FF}| \gg 1$ and $A_2(s) \cong A_{o,2}$ over the audio bandwidth, the expression in Eq. (5.21) can be approximated as,

$$\begin{aligned} &G_{FF,a}(s) \cong G_{FF,i}\left(1 - \frac{1}{A_2(s)\cdot\beta_{FF}}\right) \\ &\cong -\left(\frac{R_5}{R_6}\right)\left(1\pm\alpha_1\right)\left(1 - \frac{1}{A_{o,2}}\left[1 + \left(\frac{R_5}{R_6}\right)\left(1\pm\alpha_1\right)\left(1+\frac{R_6}{R_7}\right)\right]\right) \\ &\cong G_{FF,i}(1\pm\alpha_1) - G_{FF,i}\left(\frac{1\pm\alpha_1}{A_{o,2}}\right) \\ &\quad + (G_{FF,i})^2\frac{(1\pm\alpha_1)^2}{A_{o,2}}\left(1+\frac{R_6}{R_7}\right). \end{aligned} \tag{5.22}$$

From Eq. (5.22), it can be observed that there are multiple solutions for the two variable equation which is in the generic form $G_{FF,a} = k_1 + k_2\, x + k_3\, y + k_4\, xy + k_5\, x^2 y$, where $x = \alpha_1$ and $y = 1/A_{o,2}$. Therefore, the amount of PSRR improvement achieved by the proposed FFPSNC implementation is,

$$\text{PSRR}_{dB,imp} = -20 \cdot \log(|\alpha|) = -20 \cdot \log\left(\left|\frac{G_{FF,a} - G_{FF,i}}{G_{FF,i}}\right|\right). \quad (5.23)$$

5.4.1 *Implementation tradeoffs*

To illustrate the design tradeoffs for the implementation on the PSRR improvement, multiple solutions of Eq. (5.23) were drawn in a 3D plot shown in Fig. 5.9, where the two variables ($x = \alpha_1, y = 1/A_{o,2}$) were swept over a wide range of values. The contour plot is shown in Fig. 5.10.

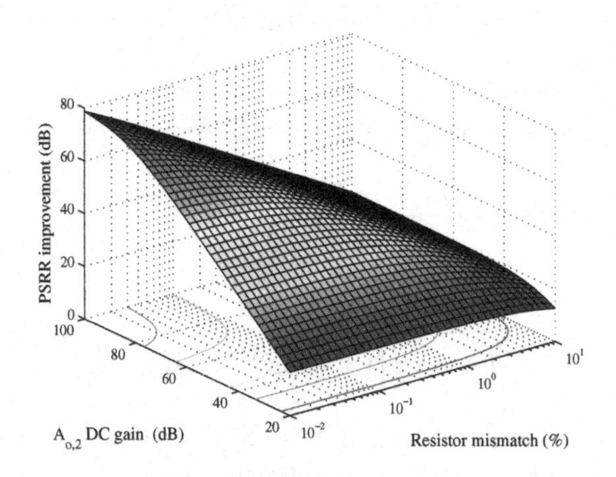

Fig. 5.9 3D-surface plot of design tradeoffs for G_{FF} implementation.

It can be observed from Fig. 5.10, that the resistor mismatch (α_1) is the dominant error parameter in Eq. (5.22) when the DC open-loop gain $A_{o,2}$ is higher than 60 dB. On the other hand, if the resistor mismatch is less than 0.02%, then $A_{o,2}$ needs to be higher than 80 dB or it will become the limiting factor in the PSRR improvement. It can be noted that the α_1 error in Eq. (5.22) depends only on the mismatch between resistors R_5/R_6. This mismatch can be minimized using less silicon area overhead compared to the area needed to minimize the error δ in the BTL CDA architecture. Another possibility to reduce the mismatch in the implementation of G_{FF} is to use

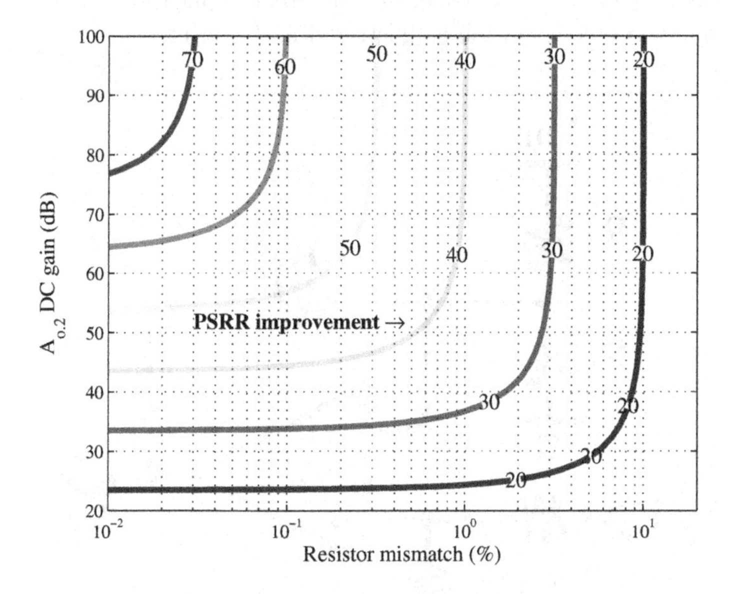

Fig. 5.10 Contour plot of design tradeoffs for G_{FF} implementation.

dynamic element matching techniques in the resistors by choosing different but almost equal-valued resistors to represent a more accurate value as a function of time [61, 122]. The goal is to transform the accuracy error due to the resistor's mismatch from a DC offset into an AC signal of equivalent power that can be removed by the noise shaping action of the compensator in the close loop system. However, this increases the complexity, area, and power consumption [122].

To illustrate the design methodology, the FFPSNC technique is implemented with $A_{o,2} = 54\ dB$, and the resistors are implemented with an expected mismatch of less than 2% (estimated from technology model document) to obtain a PSRR improvement around 34 dB, according to Eq. (5.23) and Fig. 5.10. Amplifier A_2 is implemented as a two-stage miller-compensated amplifier with DC gain of 54 dB, phase margin of 72°, and GBW of 10 MHz. It consumes a quiescent current of 27.5 μA.

5.4.2 *Design procedure*

Using the contours in Fig. 5.10, we can determine both, the minimum amount of resistor mismatch α_1 and the minimum $A_{o,2}$ gain needed for a desired value of PSRR improvement for different applications. The FFP-SNC technique implementation is illustrated in Fig. 5.11.

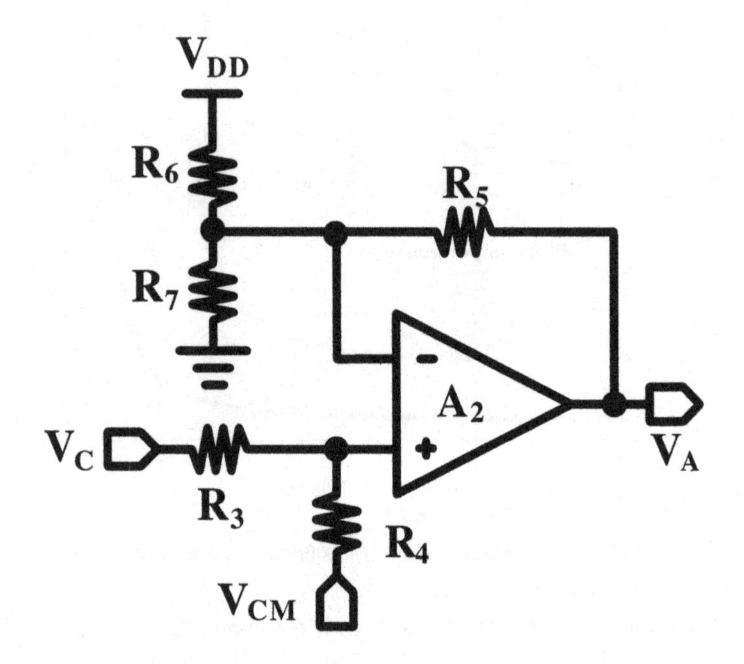

Fig. 5.11 FFPSNC technique implementation.

Table 5.1 FFPSNC technique design procedure.

	Design procedure based on Figs. 5.10 and 5.11
1.	Determine $G_{FF} = D/G_M$.
2.	Select desired PSRR improvement contour line.
3.	Find minimum $A_{o,2}$ and resistor mismatch α_1.
4.	Design amplifier A_2 for desired DC gain $A_{o,2}$.
5.	Choose resistor width from technology data for minimum α_1.
6.	Choose R_5 much larger than R_{out} of amplifier A_2 to avoid limiting the DC gain.
7.	Choose $R_6 = R_5/G_{FF}$ to implement desired FF gain.
8.	Choose $R_7 = R_6 \cdot (V_{DD} - V_{CM})/V_{CM}$ for desired V_{CM}.
9.	Choose $R_3 = R_5$ for unity gain for audio signal
10.	Choose $R_4 = R_6//R_7$ for fully balanced amplifier.

A comprehensive design procedure for the implementation of G_{FF} for

PWM is summarized in Table 5.1. For this example with a $G_M = 2$ and D=0.5, a $G_{FF} = 1/4$ will be needed. If we desire a 40 dB PSRR improvement, we can tolerate a minimum resistor mismatch of 1% with the $A_{o,2}$ gain of 70 dB. On the other hand, if the design application only needs an extra 20 dB of PSRR improvement, then we could tolerate up to 10% of resistor mismatch with an A_2 open-loop gain of 50 dB. For this example, a 30 dB PSRR improvement will be target, requiring $A_{o,2} = 50dB$ and $\alpha_1 = 2\%$. Then, after designing the amplifier, we can look that the CMOS 0.18 μm technology model document which states that a minimum resistor width of 1 μm is required for an $\alpha_1 \cong 2\%$.

The next step is to choose the resistor values. R_5 is chosen as 50 kΩ to avoid loading the amplifier output. R_6 is chosen as 200 kΩ to implement the desired $G_{FF} = 1/4$. Then, for a $V_{CM} = V_{DD}/2$, $R_7 = R_6 = 200k\Omega$. The final steps are choosing $R_3 = R_5 = 50k\Omega$ and $R_4 = R_7//R_6 = 100k\Omega$ to achieve a unity gain for the controller output containing the audio signal, and to have a fully balanced amplifier less sensitive to common mode noise. All resistors were implemented using P+ poly material over N-well with a width of 2 μm.

5.5 Simulation results

To verify the versatility of the FFPSNC technique in single-ended architectures, a first-order and a second-order PWM CDA systems were designed with and without the proposed FFPSNC technique for comparison; the MATLAB © Simulink simulation models are illustrated in Fig. 5.12 for the first order compensator and in Fig. 5.13 for the second order compensator, where $Ki = 2\pi \cdot f_{int}$ and $K_Z = 1.5$.

It can be observed that the PWM modulator plus the class-D output stage are linearized and represented by a single gain element. Also, the G_{FF} gain is implemented by a gain element with a value equal to the desired gain plus some error.

The ideal models demonstrate the basic principle behind the FFPSNC technique and can be used to simulate the supply-noise rejection using a linear analysis. However, in the real implementation the system could present some inaccuracies in the cancellation path due to its non-linear nature.

To verify that the FFPSNC technique is still valid on the transistor level design, the first order and second order PWM loops were implemented with and without the FFPSNC technique. Both circuit designs were simulated

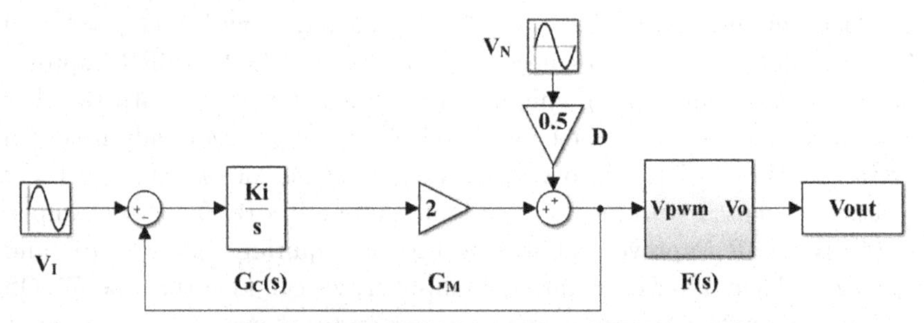

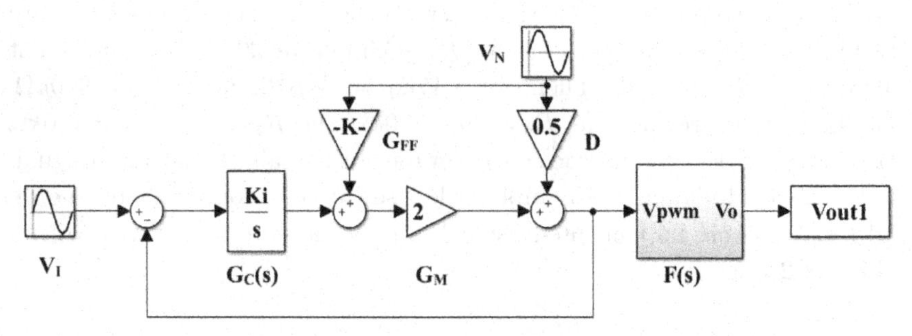

Fig. 5.12 Simulink models for 1st order PWM CDA for supply noise.

using the periodic-state analysis together with the periodic stability analysis; the simulation results are shown in Fig. 5.14. A 2% mismatch was introduced in the FFPSNC implementation by adjusting the value of the feedforward gain, and both first and second order systems achieved around 34 dB of PSRR enhancement. The FFPSNC technique could be applied to a high-order loop to increase its PSRR if it is required by the application. The additional PSRR is independent of the order of the compensator since it only depends on the accuracy of the implementation of G_{FF}.

The added cost of the second order loop is an extra integrator and the required compensation for stability of the loop. The only cost of the FFPSNC is an extra amplifier and the G_{FF} implementation, but no extra compensation is required. Moreover, the first order system with FFPSNC technique provides a better PSRR at high frequencies compared with the

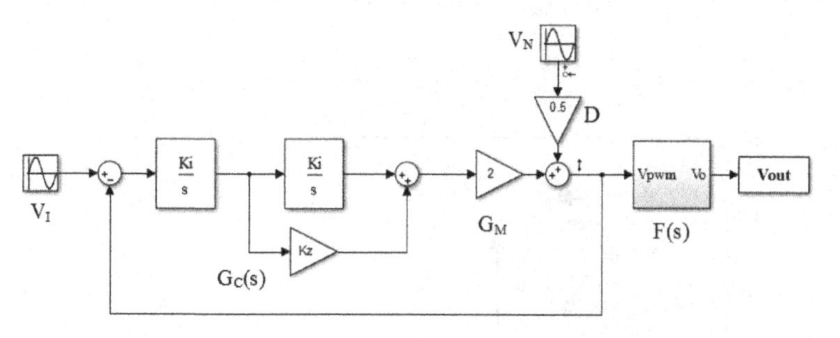

2nd Order PWM CDA without FFPSNC technique

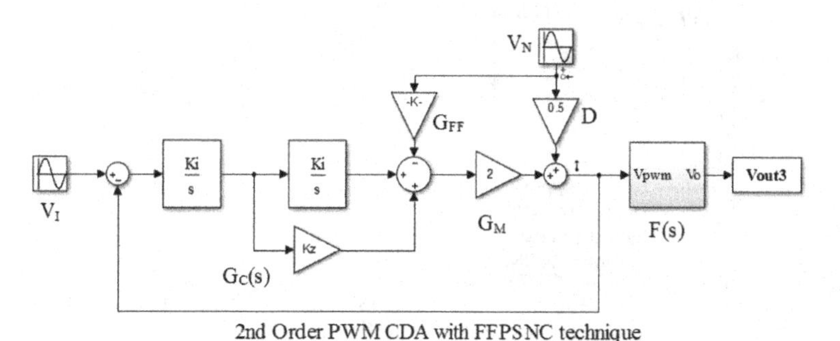

2nd Order PWM CDA with FFPSNC technique

Fig. 5.13 Simulink models for 2nd order PWM CDA for supply noise.

second order system without the technique.

To verify the robustness of the FFPSNC technique implementation, a Monte Carlo simulation with 200 runs for the first order CDA with PSRR improvement is shown in Fig. 5.15. It can be observed that the mean value for the PSRR improvement is 33.21 dB, while the standard deviation is only 2.29 dB. This is mainly limited by the resistor mismatch in the proposed implementation which is around 2%. This mismatch can be reduced by occupying more silicon area; for example, for this particular 180nm technology using P+ polysilicon resistors, to obtain a PSRR improvement of 50 dB a mismatch of 0.1 % would be needed, increasing the silicon area occupied by 220%.

5.6 Experimental results with FFPSNC technique

A first-order PWM CDA with the FFPSNC technique was fabricated in 0.18 μm CMOS standard technology, and tested with a System One Dual-

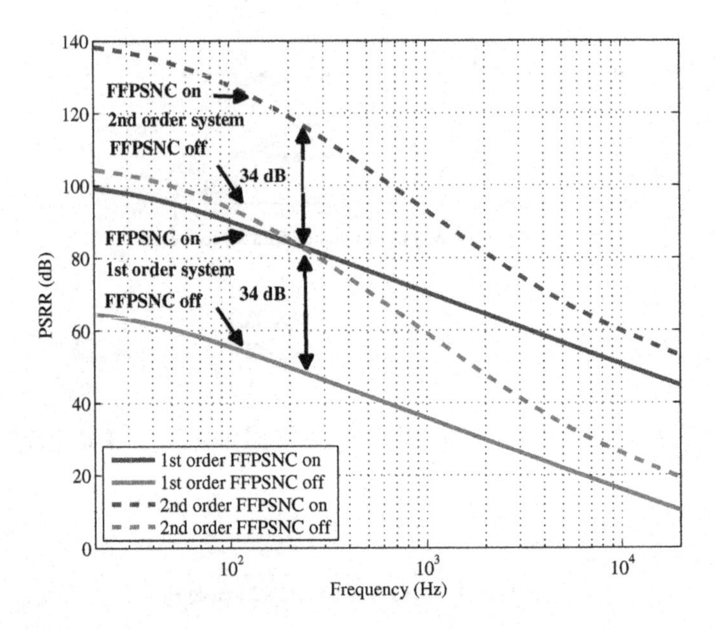

Fig. 5.14 PSRR simulation results comparison for transistor level designs.

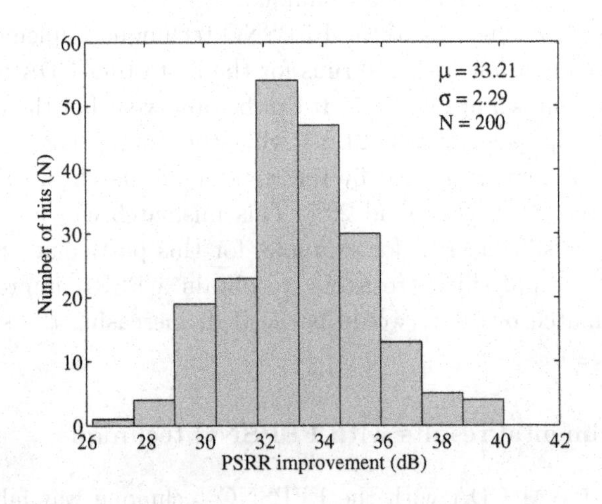

Fig. 5.15 PSRR improvement Monte Carlo simulation results.

Domain Audio Precision instrument using a 1.8 V supply voltage. The chip was encapsulated in a QFN-24 package. Figure 5.16 shows the die micrograph of the fabricated CDA, where blocks I, II, III and IV correspond to the compensator, FFPSNC technique, comparator, and output power stage, respectively. The total active area occupied by the class-D amplifier with the FFPSNC technique is 0.121 mm^2. To be able to quantify the PSRR improvement, a similar CDA was fabricated without the proposed technique.

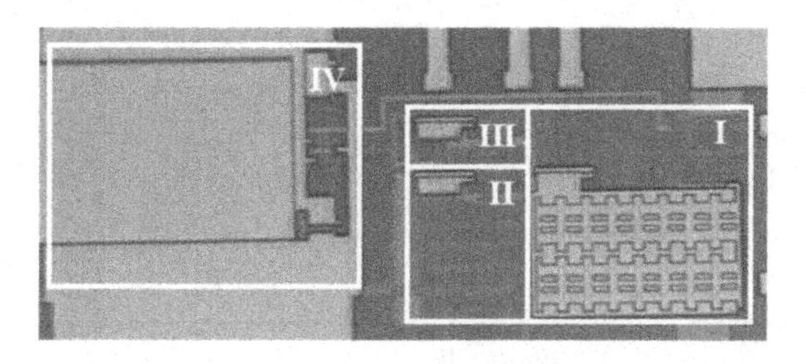

Fig. 5.16 Class-D audio amplifier die micrograph, I compensator (0.044 mm^2), II FF-PSNC technique (0.019 mm^2), III comparator (0.003 mm^2), and IV output power-stage (0.055 mm^2).

The PSRR was measured for comparison, as depicted in Fig. 1.11. Figure 5.17 shows the measured PSRR for both CDAs with the proposed FFPSNC technique and the conventional CDA without the FFPSNC technique. A peak PSRR value of 83 dB was obtained in the CDA with the proposed technique, when a 217 Hz sine-wave ripple of 250 mV was superimposed on the power-supply voltage, and no input signal was present.

From Fig. 5.17, it can be observed that the proposed technique achieves a PSRR improvement of 33 dB when compared with the similar CDA without it. This is expected from the implementation simulation shown in Fig. 5.15. Also, the FFPSNC technique is effective across the entire audio bandwidth because it doesn't affect the CDA TF, keeping the same 20 dB/dec roll-off in the frequency of interest.

It is worth noting that the FFPSNC technique can be applied to any single-ended CDA architecture to improve the PSRR performance, and that larger PSRR improvements can be achieved at the cost of additional silicon area and/or extra power consumption to improve the matching between R_5/R_6 or to achieve higher $A_{o,2}$ gain.

Table 5.2 summarizes the measured PSRR performance between the CDA with the proposed FFPSNC technique and the conventional CDA

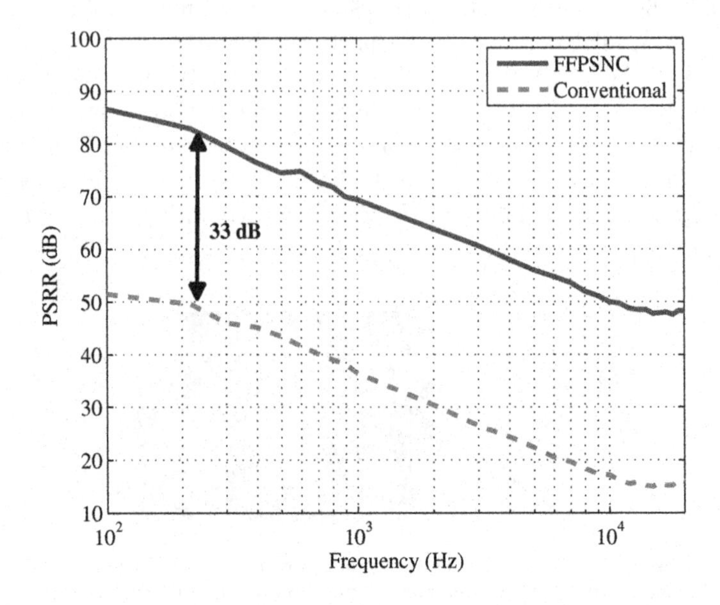

Fig. 5.17 Class-D audio amplifier PSRR measurement results.

without it. As can be seen, the proposed technique is effective across the entire audio bandwidth while adding minimum area and quiescent power to the conventional design.

Table 5.2 Comparison between conventional and FFPSNC technique.

Parameter	FFPSNC	Conventional	Difference
PSRR @ 217 Hz	83 dB	50 dB	33 dB
PSRR @ 1 kHz	69 dB	36 dB	33 dB
PSRR @ 10 kHz	50 dB	17 dB	33 dB
Active area	0.121 mm^2	0.102 mm^2	0.019 mm^2
Quiescent power	356 μW	306.5 μW	49.5 μW

The additional silicon area and quiescent power consumption are 0.019 mm^2 (16%) and 49.5 μW (14%), respectively. The additional power and area is mainly due to the implementation of G_{FF}. Also, the measured PSRR improvement is similar to the expected results from the first-order CDA simulation, as shown in Figs. 5.14 and 5.15.

The proposed FFPSNC technique does not affect the CDA's loop parameters; thus, the following measurements achieved the same results with and without the proposed technique. The output spectrum of the system with an input $V_I = 0.5\ V_{RMS}$ at 1 kHz is illustrated in Fig. 5.18. The difference between the fundamental tone and the largest harmonic is -76.5 dB.

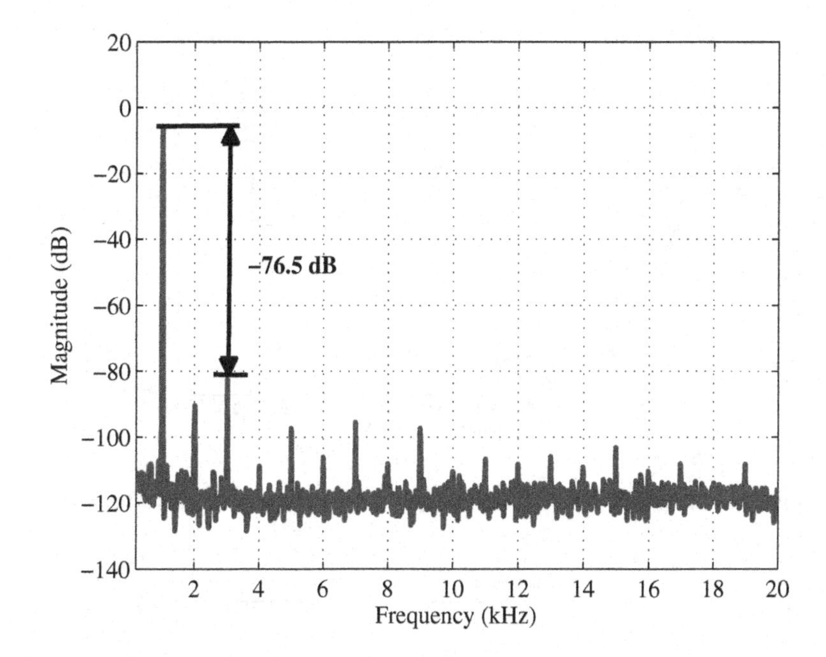

Fig. 5.18 Measured output spectrum for CDA with FFPSNC technique with Vin= 0.5 Vrms at 1 KHz.

The measured THD+N versus output power is shown in Fig. 5.19. A minimum THD+N of 0.0149% was measured in the CDA prototype, and a minimum SNR of 84 dB was measured across all the audio bandwidth, as observed in Fig. 5.20.

Figure 1.16 shows the efficiency measurement setup where an 8 Ω resistor is used as the load, and low-value sensing resistors are used in series with the supply voltage and the load to measure the input and output current, respectively. Figure 5.21 shows the measured amplifier efficiency versus the output power range from 2 mW to 250 mW.

A maximum efficiency of 94.6% was measured when delivering 150 mW of output power. Since the output stage was optimized for operation in the low to medium output power range, the efficiency curve has its peak in the

Region II of the efficiency curve, as discussed in Chapter 2 and Fig. 2.4.

The power-supply intermodulation distortion (PS-IMD) provides a lin-

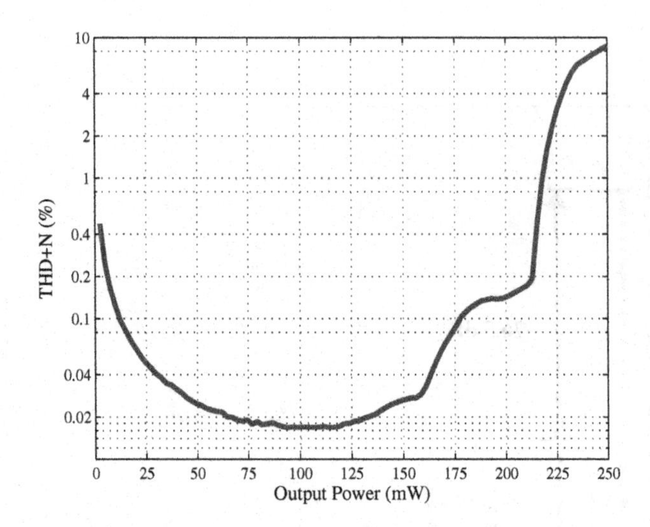

Fig. 5.19 Measured THD+N versus output power for CDA with FFPSNC technique.

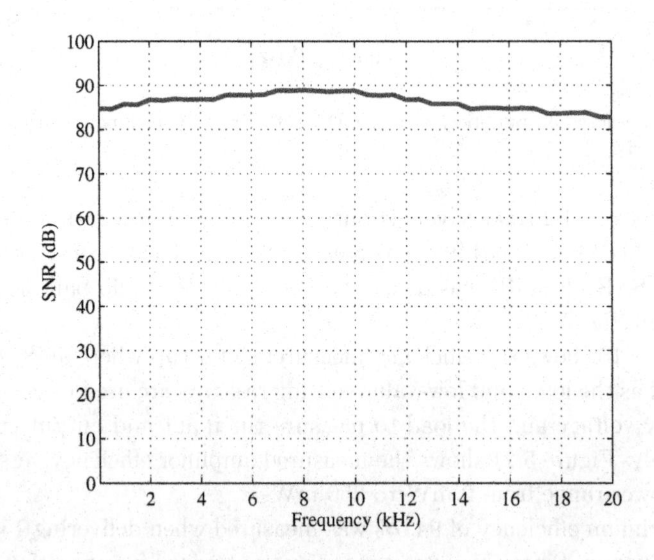

Fig. 5.20 Measured SNR for proposed CDA with FFPSNC technique.

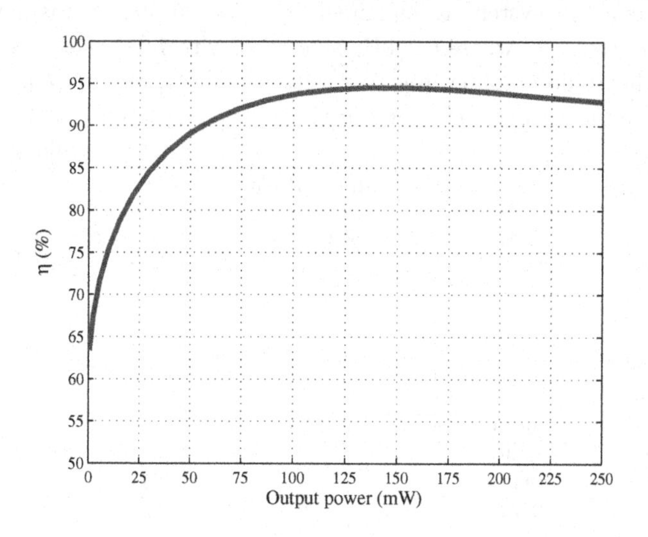

Fig. 5.21 Measured efficiency versus output power for CDA with FFPSNC technique.

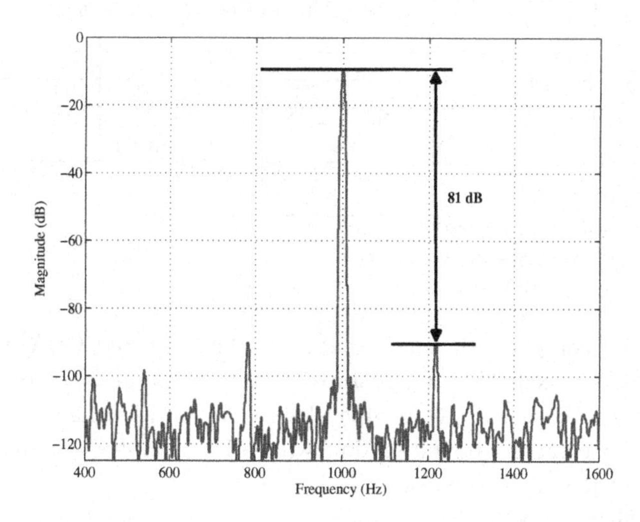

Fig. 5.22 Power-supply intermodulation distortion measurement for CDA with FFP-SNC technique.

earity metric to quantify the intermodulation distortion between the amplifier's power-supply noise and the input audio signal, when both signals are present in the system as explained in [44, 56, 75, 79]. It was measured as show in Fig. 1.13 where a 1 kHz sine wave with 1 V_{pp} was used as the input of the audio amplifier together with 0.1 V_{pp} at 217 Hz signal at the amplifier supply source. The measured frequency spectrum of the output signal is shown in Fig. 5.22. As can be seen, both intermodulation tones are at least 81 dB below the fundamental tone.

Table 5.3 Comparison of FFPSNC technique with state-of-the-art.

Parameter	FFPSNC technique [71][a]	[56]	[72]	[77]	[118]	[44]	[79]
Compensator order	1	1	3	7	4	2	1
PSRR(dB) @ 217Hz	83	70	88	65	82	96	82
I_Q(mA)	**0.20**	4.70	3.02	22.00	1.40	4.00	0.55
P_Q(mW)	**0.36**	14.98	11.17	194.00	3.50	10.00	1.49
η(%)	**94.6**	75.5	85.5	88	80	93	84
Area (mm^2)	**0.121**	0.44	1.01	10.15	0.30	1.44	1.65
Process (μm)	0.18 CMOS	0.09 DMOS	0.18 CMOS	0.6 BCD MOS	0.065 CMOS	0.25 CMOS	0.5 CMOS
THD+N(%) @ 1 kHz	0.0149	0.0300	0.0180	0.0012	0.0132	0.0012	0.0200
Supply(V)	1.8	4.2	3.7	5.0	2.5	2.5	2.7
F_{SW}(kHz)	500	410	320	450	667	1000	380
Max P_{out}(W) @ 8Ω load	0.25	0.70	1.15	10 (6Ω)	0.05 (32Ω)	3.60	0.41
Load configuration	SE	BTL	BTL	BTL	SE	BTL	BTL

SE: Single-Ended, BTL: Bridge-Tied Load.
[a]Do not include triangle-wave generator.

Table 5.3 compares the performance of the presented CDA with the FFPSNC technique to that of the state-of-the-art. As can be seen, the fabricated CDA achieves a PSRR comparable to CDAs with BTL architectures or high-order compensator filters, but with a low complexity implementation and low power consumption.

The PSRR can be improved further by decreasing the resistor mismatch in the G_{FF} implementation. This can be accomplished by occupying more silicon area using large widths in the resistor's layout; for this particular technology, a 0.1% can be achieved without special trimming by using a 30 μ width in the resistor. Also, dynamic element matching techniques could be implemented to achieve high accuracy in the resistor ratio but this would increase drastically the complexity as well as the silicon area occupied.

The PSRR performance of the single-ended can be enhanced further if the FFPSNC technique is applied to a high-order loop. For example, the second order compensator would use double the area and power of the first order compensator but the PSRR can achieve more than 100 dB at 217 Hz if the FFPSNC technique is applied, as observed in Fig. 5.14. The added cost to the actual design would be an additional 0.044 mm^2 (36%) area occupied and 51 μW (14%) power consumption.

5.7 Final remarks

The design methodology, implementation, and tradeoffs of a feed-forward power-supply noise cancellation technique were clearly delineated in this Chapter; the proposed technique is capable of achieving high PSRR in single-ended class-D audio amplifiers. The attractive features of this approach are its simplicity and effectiveness. The tradeoffs for its utilization in several applications were discussed. A first-order single-ended PWM class-D audio amplifier was fabricated to demonstrate the effectiveness of the proposed technique. The class-D amplifier prototype achieves a PSRR of 83 dB at 217 Hz, a THD+N of 0.0149%, and a maximum efficiency of 94.6%. The proposed technique enhances the fabricated CDA's PSRR by 33 dB across the entire audio bandwidth compared with a conventional CDA without it. The class-D audio amplifier prototype was implemented using 0.18 μm CMOS standard technology and occupies a total area of 0.121 mm^2. It consumes a total of 356 μW of quiescent power.

Chapter 6

Sliding-Mode Control for Class-D Amplifiers

6.1 Motivation for non-linear controllers

Due to their ideally perfect efficiency and linearity, class-D amplifiers have become a very attractive solution to implement audio drivers in applications with crucial power consumption and low-voltage requirements. However, component non-idealities degrade the audio quality when compared to the ideal amplifier. Hence, in order to achieve low distortion, the audio modulator often becomes complex and power hungry. The sliding mode control (SMC) technique is applied to the class-D audio power amplifier. The main benefit of SMC is that it avoids the triangular wave signal used in conventional class-D audio amplifiers, reducing the power consumption.

Sliding mode theory starts its development in the 1950s as an alternative solution for control problems in systems with discontinuous differential equations. It is mostly applied to variable structure systems (VSS) where each one of their subsystems is continuous although not necessarily stable. Systems with SMC include robotics, aircraft control, power converters, PWM control, and remote vehicle control [92]. This chapter presents three architectures and two methodologies to implement high performance class-D audio amplifiers with minimal power consumption [78–82]. The first two topologies, a binary modulation amplifier (BMA) and a ternary modulation amplifier (TMA), are based on the same principle of a hysteretic controller with SMC. The last class-D amplifier is implemented with integral sliding mode which provides higher loop gain across the audio frequency band, and avoids the use of differentiators which are prone to high frequency noise.

6.2 Class-D amplifier with sliding mode controller

The class-D audio amplifier with SMC conceptual diagram is shown in Fig. 6.1, where V_I, V_{PWM}, and V_O are the input (reference) audio signal, the pulse-width modulated waveform, and the output signal, respectively. It consists of four basic subsystems: the controller, a hysteresis comparator, the output power stage, and the output filter.

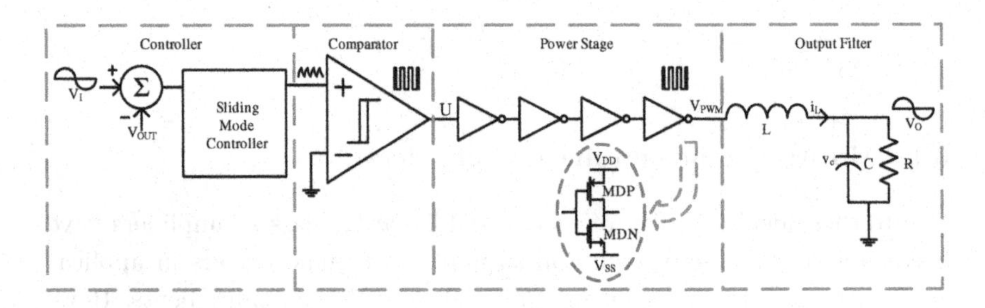

Fig. 6.1 Conceptual diagram of proposed class-D audio amplifier with SMC.

The controller and the comparator, which generate the pulse-width modulated signal by using sliding mode control, are both integrated in a single-chip along with the power stage. The output filter is designed with a cutoff frequency of 20 kHz using off-chip components due to its large size. A Butterworth filter approximation is chosen due to its flat frequency response. Note that the feedback signal includes the output filter, allowing the closed loop to correct for errors in the output filter components, as discussed in Chapter 3.

The sliding mode controller design is based on the state variables of the system to be controlled. For this particular case the system consists of the low-pass RLC filter placed at the end of the class-D audio amplifier.

If we consider just the last inverter of the power stage, the circuit shown in Fig. 6.2 is obtained. In this figure, the two different substructures during the class-D audio amplifier operation can be observed. In the first part of the cycle, depicted as Fig. 6.2(a), transistor MDP is ON and transistor MDN is OFF, that is, the input V_{PWM} equals to V_{DD}. For the second subinterval, Fig. 6.2(b), transistor MDP is OFF and transistor MDN is ON, i.e. the input V_{PWM} is equal to V_{SS}. Then, the dynamical state equation of the low-pass filter at the output of the class-D audio amplifier is given by

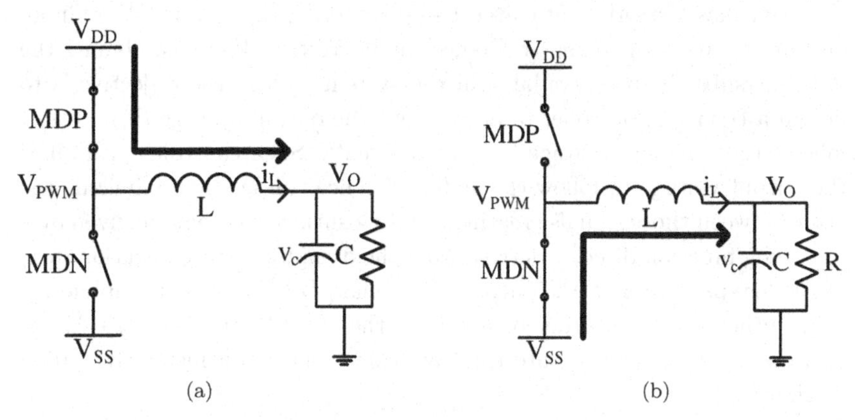

Fig. 6.2 Subintervals of operation in class-D audio power amplifier under sliding mode control (a) Subinterval I and (b) Subinterval II.

$$\begin{pmatrix} \frac{d}{dt} i_L \\ \frac{d}{dt} v_C \end{pmatrix} = \begin{pmatrix} 0 & -\frac{1}{L} \\ \frac{1}{C} & -\frac{1}{CR} \end{pmatrix} \begin{pmatrix} i_L \\ v_C \end{pmatrix} + \begin{pmatrix} \frac{1}{L} \\ 0 \end{pmatrix} V_{PWM} \tag{6.1}$$

where the state variables i_L and v_C denote the inductor current and the capacitor voltage, and V_{PWM} is the input signal that can be either V_{DD} or V_{SS}.

The low-pass filter is a second-order stable system with negative and imaginary eigenvalues that yields a stable focus natural equilibrium point [92] for each case (V_{DD} or V_{SS}). Depending on which part of the cycle is operating, the response of the low-pass filter would be that of the value of V_{PWM}. Then, we would have two different phase portraits, i.e. a plot of typical trajectories of the state variables v_C and i_L in the state space system defined in equation (6.1) for different initial conditions, each one corresponding to the values of the signal V_{PWM}, either V_{DD} or V_{SS}.

Figure 6.3(a) shows the step response of the low-pass filter in the configuration shown in the subinterval I in Fig. 6.2(a). Notice that the value of the output voltage (capacitor voltage v_C) goes to the positive supply voltage after a short transient. On the other hand, Fig. 6.3(b) illustrates the case when a step is applied to the low-pass filter configured as shown in Fig. 6.2(b). In this case, the value of the capacitor voltage v_C goes to the negative voltage supply. Observe that, at steady state, the value of the current across the inductor i_L is proportional to the value of the normalized resistor R.

Even though the nature of the low-pass filter is asymptotically stable, i.e. it reaches a steady state after a step response, the goal is to obtain an output signal equal to the (reference) audio voltage V_I by combining the different substructures available in the system. Thus, the objective is to design a tracking controller to ensure that the output voltage ($V_O = v_C$) follows the reference voltage V_I (audio signal). Such controller will allow the output voltage to follow the audio reference voltage by minimizing the error between those signals creating a sliding surface that will be given by a switching function directly derived from the dynamical state equation (6.1) of the low-pass filter at the output of the class-D audio power amplifier.

In other words, in order to minimize the error $V_{e1}(t) = V_I(t) - V_O$, it is necessary to design a state feedback control law to achieve asymptotic tracking.

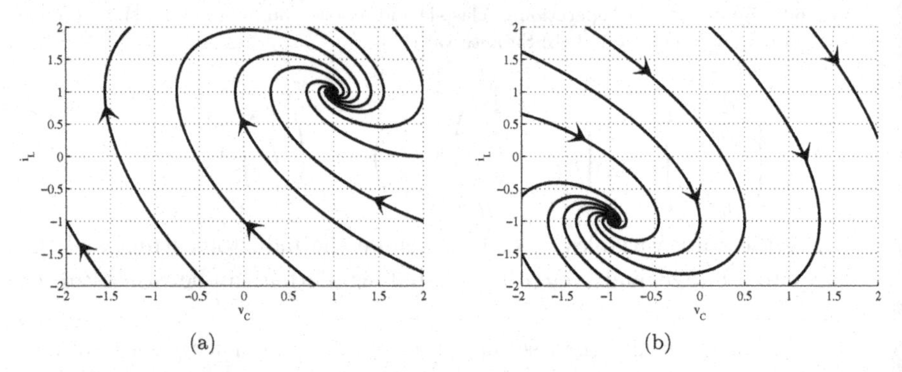

Fig. 6.3 Normalized phase portraits of subintervals I and II: (a) $V_{PWM} = V_{DD}$ and (b) $V_{PWM} = V_{SS}$.

The class-D audio amplifier operating under sliding mode control consists of two different parts. The first part corresponds to the so called reaching mode, i.e. from any initial condition; the system will reach the sliding surface. Once there, the second part is the motion from the sliding surface to the equilibrium point of the system, i.e. the sliding mode.

The sliding mode controller will make the system to switch between V_{DD} and V_{SS} according to the sign of the switching function.

$$V_I = \begin{cases} V_{DD} & \text{when } s(V_{e1}, t) > 0, \\ V_{SS} & \text{when } s(V_{e1}, t) < 0. \end{cases} \tag{6.2}$$

The controller makes the system to satisfy the reaching condition and, on the other hand, the fact that the sliding equilibrium point [94] of the class-D audio amplifier is a stable node with eigenvalues real and negative, as derived in Appendix B, guarantees the sliding mode of the system toward its sliding equilibrium point. The sliding mode controller makes the class-D amplifier a stable system with a stable node equilibrium point where any initial point in the phase portrait reaches the sliding surface and then moves to the sliding equilibrium point of the system, as shown in Fig. 6.4.

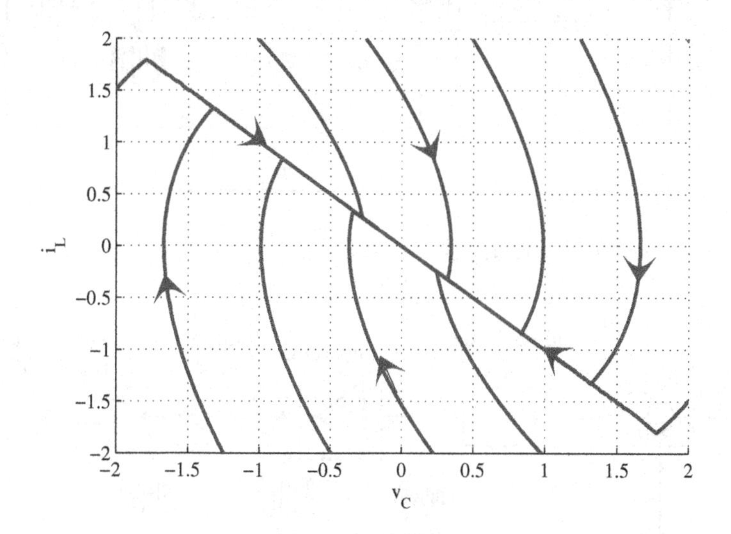

Fig. 6.4 Normalized sliding mode operation in class-D audio power amplifier.

6.2.1 *Controller design with linearity enhancement*

The class-D audio amplifiers are based on the block diagram shown in Fig. 6.5. The architecture implements a sliding mode controller (SMC) defined by the control law, or switching function (SF)

$$s(V_{e1}, t) = V_{e1}(t) + \alpha \dot{V_{e1}}(t) = V_{e1}(t) \;+\; \alpha V_{e2}(t) \qquad (6.3)$$

where the sliding parameter α is calculated to meet the Hurwitz stability criterion and to guarantee a fast and smooth transient response with flat delay characteristic, and $V_{e1}(t) = V_I(t) - V_O(t)$ is the voltage error function.

An extra local loop with negative feedback β is added to improve the linearity without jeopardizing the stability. This factor β requires an extra adder node as shown in Fig. 6.5. Figure 6.6 illustrates an equivalent but topologically simpler block diagram that eliminates one adder.

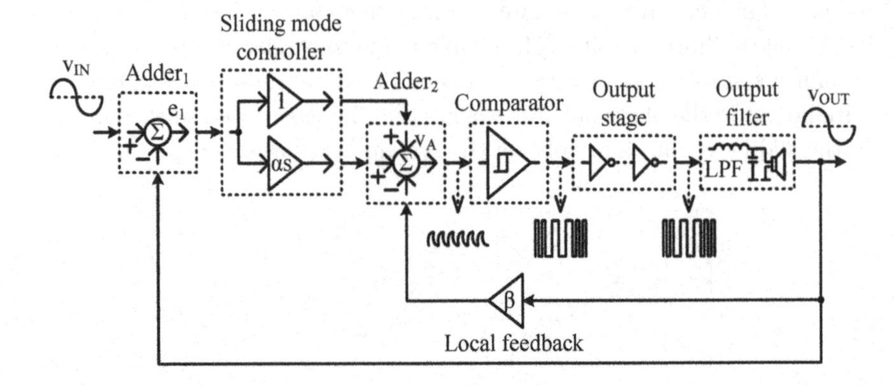

Fig. 6.5 Class-D audio power amplifier two-adder implementation.

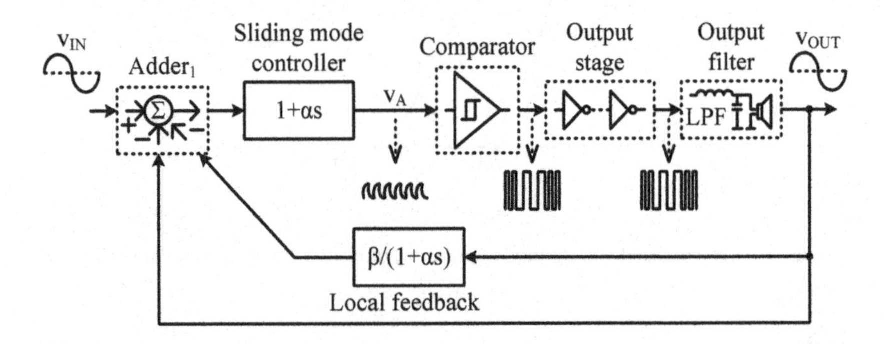

Fig. 6.6 Class-D audio power amplifier one-adder implementation.

The remaining blocks of the system in Fig. 6.5 are the hysteresis comparator and the output power stage. The comparator converts the equivalent analog control signal to a pulse-width modulated (PWM) signal, according to the sign of the switching function. The power stage generates the driving capability to supply the output current to the load, and simultaneously minimizes the output resistance. An off-chip second-order flat-response low-pass filter (LPF), with typical component values L = 45 μH, C = 1.4 μF, and an 8 Ω speaker, recovers the analog audio signal. If needed, the class-D amplifier can be converted to a filter-less architecture by using the parasitic components of the speaker as an embedded filter and adjusting the coefficient α in the controller to fit the speaker model parameters.

A noise constraint to implement the sliding mode controller is the lossless differentiator function in equation (6.3), due to its infinite bandwidth. A practical solution is to implement a lossy-differentiator (LD) function, with a finite bandwidth ω_p, in the control law as

$$S(V_{E1}, s) = \left[1 + \left(\frac{\alpha s}{\dfrac{s}{\omega_p} + 1}\right)\right] V_{E1}(s) \tag{6.4}$$

where $\omega_p = 2\pi f_p$. This lossy-differentiator, together with a finite bandwidth operational amplifier (OPAMP) bounds the class-D audio amplifier bandwidth required in the system and limits the high-frequency noise that could affect the signal-to-noise ratio (SNR) of the audio power amplifier.

Figure 6.7 illustrates two cases where it can be appreciated that the lossy differentiator with pole $\omega_{p2} = 20 / \alpha$ gives better linearity than that with $\omega_{p1} = 10 / \alpha$, since the higher-frequency pole allows for a switching function closer to the ideal control law expressed in equation (6.3). On the other hand, a lossy-differentiator with a higher-frequency pole translates into larger overall controller bandwidth and hence more quiescent power consumption. We thus have a design trade-off between linearity and power consumption of the class-D amplifier. The value $\omega_p = 10 / \alpha$ was chosen for the particular design described in this chapter.

The linearity of the class-D audio amplifiers increases because the distorting harmonics are reduced by the feedback factor β, as shown in equation (6.5). However, from the expression in equation (6.6), the fundamental tone is also attenuated.

$$THD = \sqrt{\left(\frac{HD_2}{(1+\beta)^2}\right)^2 + \left(\frac{HD_3}{(1+\beta)^3}\right)^2 + \cdots + \left(\frac{HD_N}{(1+\beta)^N}\right)^2}, \tag{6.5}$$

$$V_O = \left(\frac{1}{1 + \beta}\right) V_I. \tag{6.6}$$

Figure 6.8 illustrates the effect of the hysteresis-window width in the linearity of the class-D audio amplifiers and Fig. 6.9 shows the effect of the hysteresis voltage window in the switching frequency of the amplifiers.

The switching frequency, as it will be shown later, increases inversely proportional to the hysteresis voltage window width [55]. In an ideal system operating with sliding mode control, the hysteresis window is zero and the

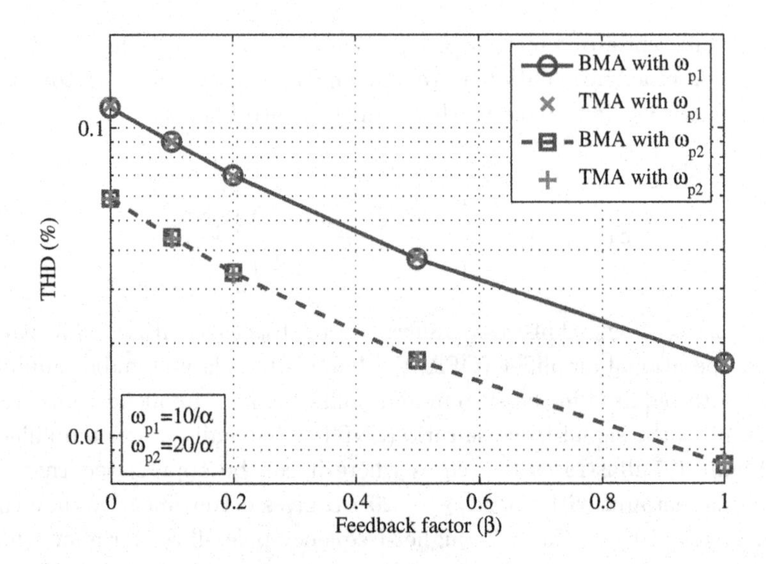

Fig. 6.7 Effect of feedback factor β in the linearity performance of the class-D audio amplifiers.

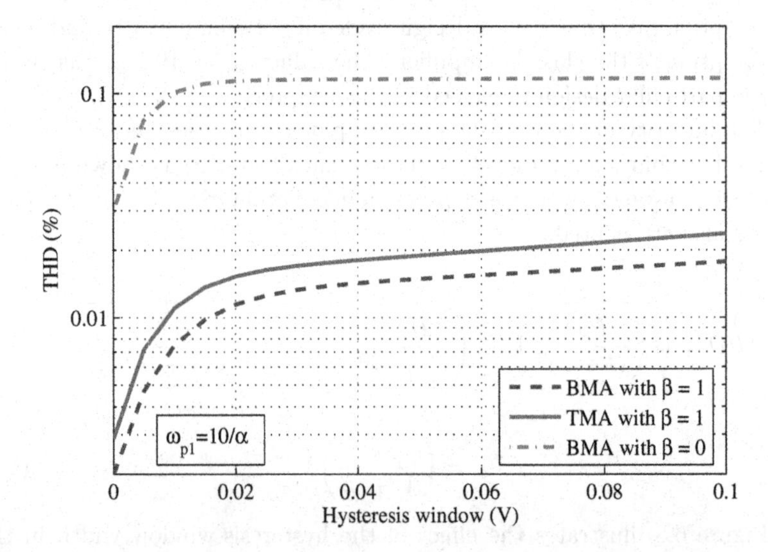

Fig. 6.8 Effect of the hysteresis-window width in the linearity performance of the class-D audio amplifiers.

switching frequency is infinite. Also notice that both amplifiers BMA/TMA provide the same linearity, provided that they switch around the same

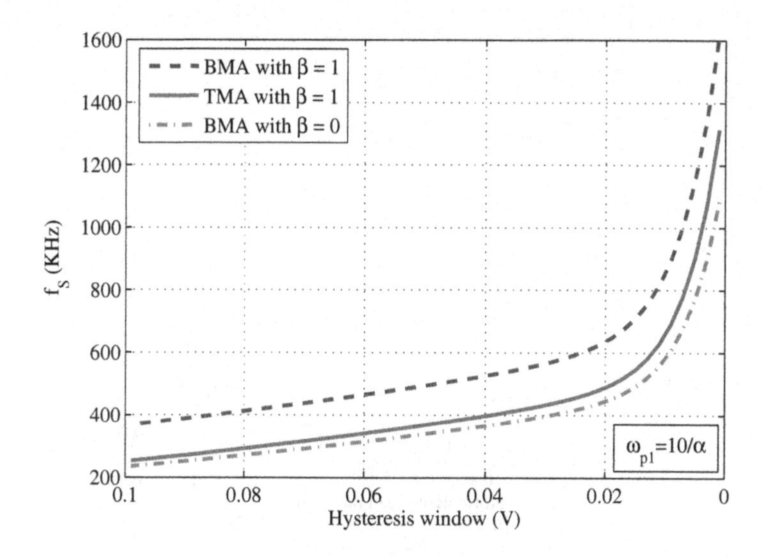

Fig. 6.9 Effect of the hysteresis-window width in the class-D audio power amplifiers switching frequency.

frequency. This fact is of particular importance because the odd carrier harmonics of the TMA cancel due to the additional modulation level, and its effective switching frequency doubles [54].

There exists a practical trade-off between the amplifier's frequency of operation and its efficiency because the dynamic power losses are proportional to the operating frequency [74]. Figure 6.10 shows the theoretical class-D amplifier efficiency when the switching frequency is increased from 200 kHz to 2 MHz. Even though the efficiency at full power is still high, the class-D audio amplifier with higher switching frequency presents significant efficiency reduction at the most common load configurations.

Finally, Fig. 6.11 shows the effect of increasing the feedback factor β in the switching frequency and in the linearity of the class-D amplifiers, when the hysteresis window is kept fixed. It can be appreciated that the switching frequency variation is small in both amplifiers.

6.2.2 *Architecture of class-D amplifiers with SMC*

The implementation of the switching function described in equation (6.4), by using finite bandwidth circuits (the operational amplifier is characterized by one single dominant pole at ω_{3dB}), modifies the control law and adds an extra pole to the system, but it does not compromise the class-D audio

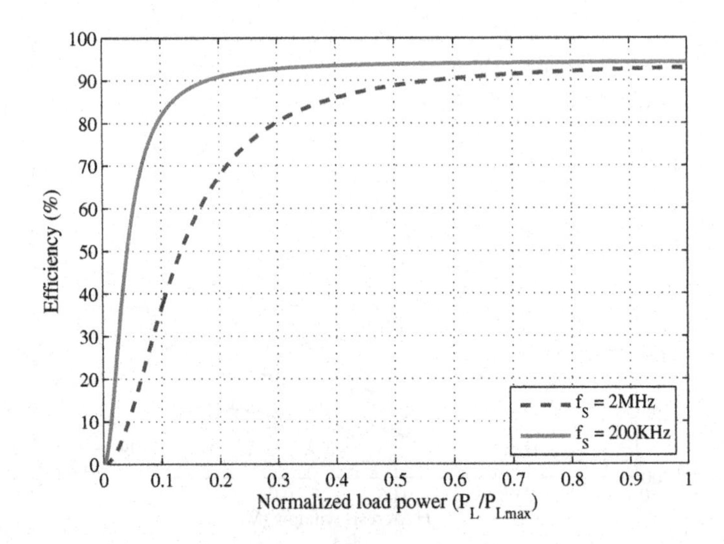

Fig. 6.10 Effect of the increment of switching frequency in the class-D audio amplifiers efficiency performance.

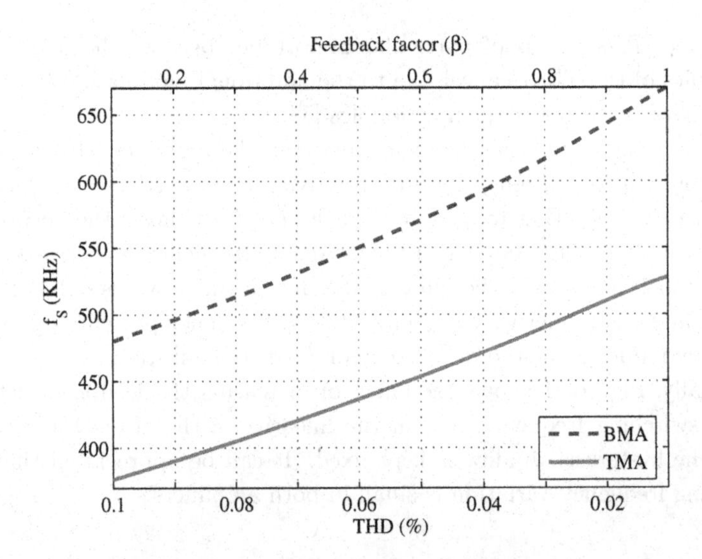

Fig. 6.11 Effect of β on class-D amplifiers switching frequency. Lower (upper) horizontal axis represents THD (β).

amplifier stability. As mentioned before, this extra pole limits the high frequency noise and bounds the class-D amplifier bandwidth. The new switching function including the additional pole is

$$S(V_{E1}, s) = \left[\frac{1}{\left(1 + \frac{s}{\omega_1}\right)} + \frac{\alpha s}{\left(1 + \frac{s}{\omega_2}\right)\left(1 + \frac{s}{\omega_3}\right)} \right] V_{E1}(s) \quad (6.7)$$

where ω_1 is the extra pole introduced by the OPAMP closed loop finite bandwidth, and ω_2 and ω_3 are the poles affected by the finite OPAMP closed loop pole (ω_{3dB}) and the lossy-differentiator pole (ω_p). Note that

$$\omega_1 > \omega_2, \omega_3. \quad (6.8)$$

The control law is built using the minimum number of components in both amplifiers in order to reduce silicon area, and more importantly, to reduce the static power consumption.

6.2.2.1 *Binary modulation amplifier*

Recalling the class-D amplifier architecture shown in Fig. 6.5, its straightforward active-RC implementation consists of three different building blocks: two adders and one lossy-differentiator, as shown in Fig. 6.12 [80]. Instead, if the class-D amplifier architecture is modified as shown in Fig. 6.5, the whole controller can be implemented using one single block if a fully-differential topology is used and since β is moved to the first adder node, the second adder can be eliminated. Therefore, the area and power consumption of the controller reduce considerably. Figure 6.13 shows the binary modulation amplifier (BMA) architecture. Examining node V_A at Fig. 6.5 yields

$$V_A(t) = \left[V_I(t) + \alpha \frac{d}{dt} V_I(t) \right] - \left[(1+\beta) V_O(t) + \alpha \frac{d}{dt} V_O(t) \right]. \quad (6.9)$$

Note that for $\beta = 0$, equation (6.9) becomes the ideal switching function expressed before in equation (6.3). In the actual active-RC implementation, neglecting the operational amplifier non-idealities, and after the ideal differentiator (αs) is replaced by a lossy one $[\alpha s \ / \ (1 + s \ / \ \omega_p)]$, $V_A(s)$ becomes

$$V_{A\pm}(s) = \left(1 + \frac{\alpha s}{1 + \frac{s}{\omega_p}}\right) V_{I\pm}(s) \pm \left(1 + \beta + \frac{\alpha s}{1 + \frac{s}{\omega_p}}\right) V_{O\mp}(s) \quad (6.10)$$

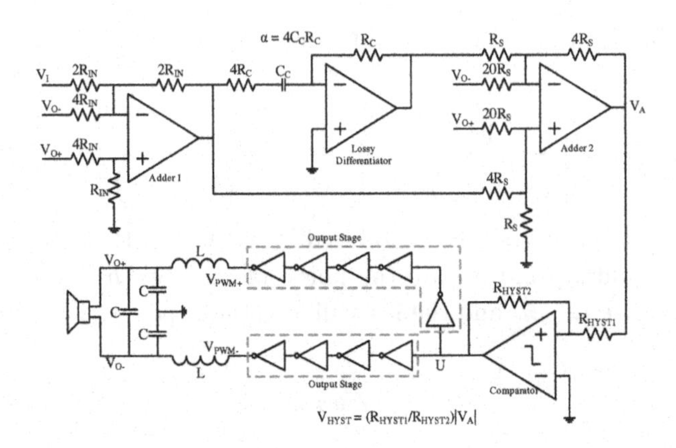

Fig. 6.12 Single-ended BMA architecture.

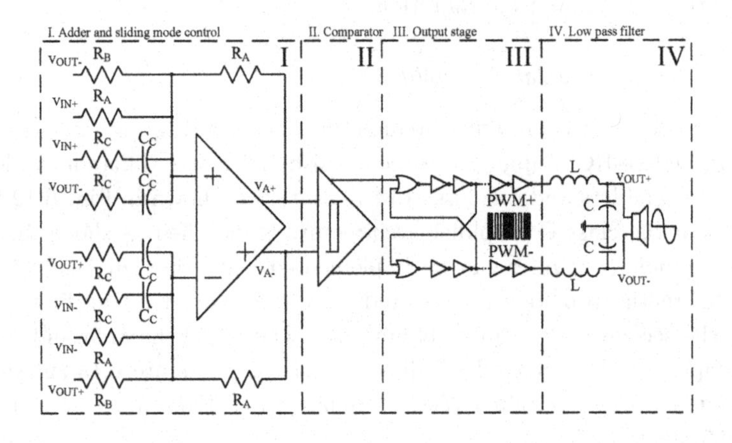

Fig. 6.13 Fully-differential BMA architecture.

where $\alpha = R_A C_C$, $\omega_p = 1 / R_C C_C$ and $(1 + \beta) = R_A / R_B$.

Hence, a single fully-differential operational amplifier implements the error function, the lossy-differentiator, and the local feedback β. The signal V_A represents the output of the sliding mode controller and the feedback factor β. The signal V_A, generated by the previous stage, is transformed into a binary signal, pulse-width modulated, which is fed to the load through the output stage. Figure 6.14 illustrates a simulated differential pulse-width modulated signal as well as the input and output voltages. Note that the output voltage is attenuated with respect to the input signal due to feedback factor β.

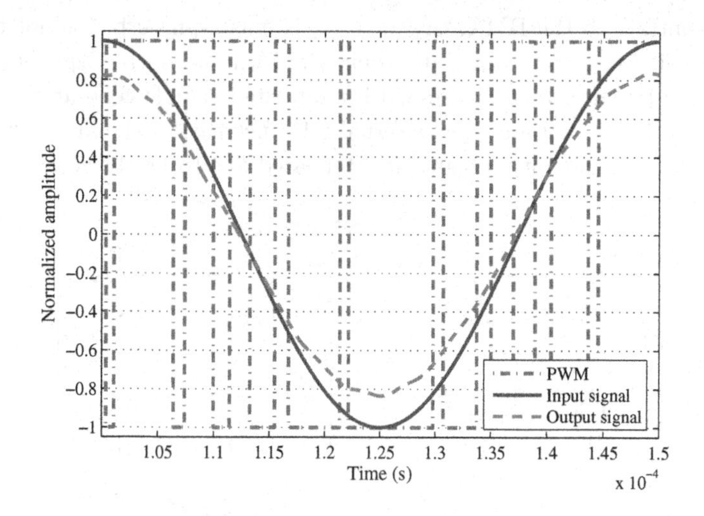

Fig. 6.14 Input and output signals in the BMA.

6.2.2.2 *Ternary modulation amplifier*

The ternary modulation amplifier (TMA), shown in Fig. 6.15, uses two single-ended operational amplifiers, and the core implementation of the switching function with the lossy-differentiator as described in equation (6.4), however its topology is based on the architecture used in conventional PWM schemes to generate ternary modulation based on a single carrier [56]. More specifically, as shown in Fig. 6.16, in traditional architectures a single ramp wave is compared to a differential analog signal, generating two binary signals that, when subtracted, generate a third modulation level.

Since the proposed TMA topology, as well as the BMA, lacks of any reference carrier signal, the SMC and β factor are implemented by

$$V_{A\pm}(s) = \left(1 + \frac{\alpha s}{1 + \frac{s}{\omega_p}}\right) V_{I\pm}(s) \mp \left(1 + \beta + \frac{\alpha s}{1 + \dfrac{s}{\omega_p}}\right) V_{O\pm}(s) \qquad (6.11)$$

However, notice that the input $V_I(s)$ and output $V_O(s)$ voltage signals create two independent single-ended loops driving a differential load, generating two different but complementary switching functions to be applied to the comparators. The difference of these two out-of-phase binary signals thus creates three voltage levels.

In contrast to the BMA, the wave V_A is out-of-phase but is not fully-differential. Figure 6.17 illustrates typical TMA input/output signals. Note that the difference of the pulse-width modulated signals generates a wave with three levels, without any external reference carrier signal. Just as in the BMA case, the output signal is attenuated due to feedback factor β.

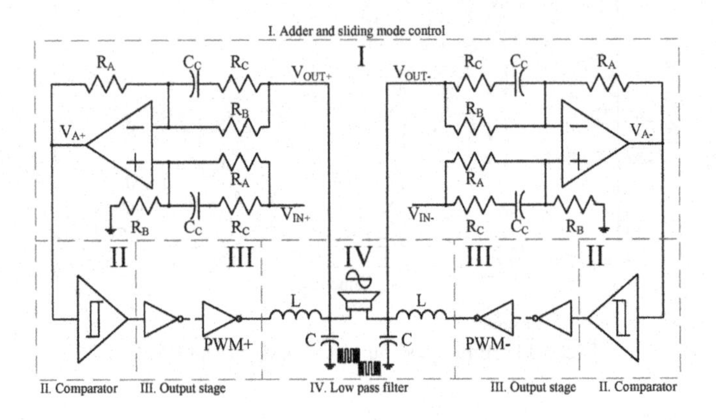

Fig. 6.15 Ternary modulation amplifier (TMA) architecture.

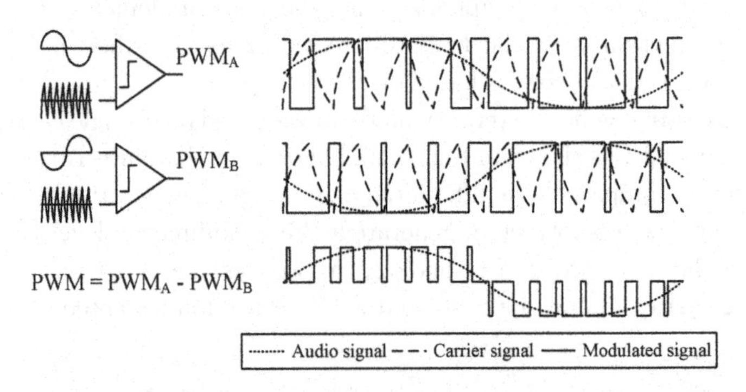

Fig. 6.16 Conventional ternary modulation scheme where the two top signals are single-ended waveforms whose difference is the bottom signal.

6.2.3 *Switching frequency with SMC*

The switching frequency of the class-D audio power amplifiers BMA and TMA is directly related to the hysteresis band in the comparator because the system will toggle between states every time it reaches the hysteresis

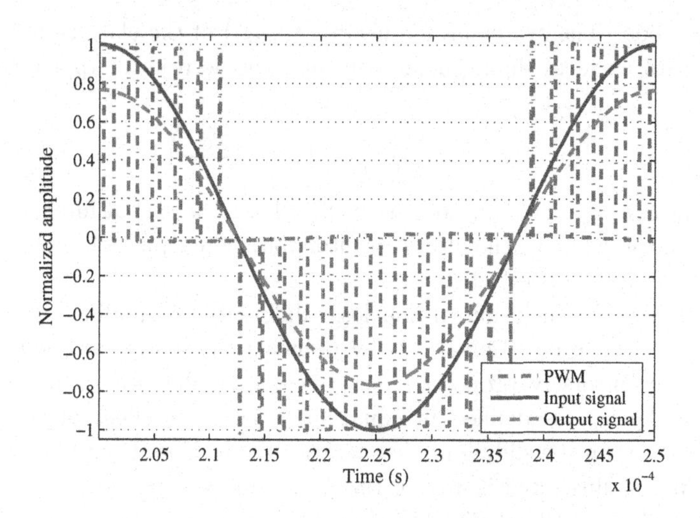

Fig. 6.17 Input and output signals in the TMA.

voltage. Figure 6.18 shows a magnified view of the ideal switching function $s(V_{e1}, t)$ when it is operating under sliding mode.

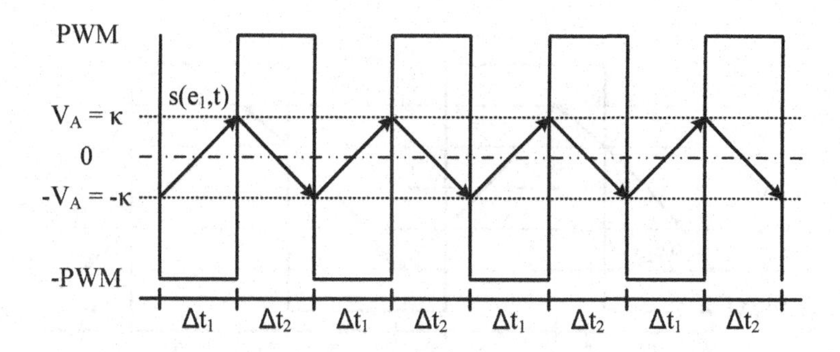

Fig. 6.18 Magnified view of ideal sliding mode operation for class-D amplifiers.

The sliding mode operation can be divided into two different subintervals of operation Δt_1 and Δt_2. During the first subinterval of operation, the voltage V_A, defined in equation (6.9), increases until it reaches the hysteresis voltage κ and the pulse-width modulated signal (PWM) goes positive. In the second subinterval, the voltage V_A decreases until its value equals the negative hysteresis voltage $-\kappa$ and then, the pulse-width modulated signal (PWM) goes negative. This cycle repeats in a steady operation during sliding mode. The switching frequency ($f_{s,ideal}$) of the class-D audio amplifier with an ideal sliding mode controller, as derived in Appendix C, is given by

$$f_{s,ideal} = \frac{1}{2\kappa}\frac{R}{L}v_C\left(1 - \frac{v_C}{V_{DD}}\right) \tag{6.12}$$

where R, L, κ, v_C, and V_{DD} are the speaker load, the output filter inductor, the hysteresis window, the filter capacitor voltage (output voltage), and the supply voltage, respectively.

However, the inclusion of the lossy-differentiator modifies the previous expression in equation (6.12) by reducing the switching frequency in an amount inversely proportional to the lossy-differentiator bandwidth, bounded by ω_p, as defined in equation (6.4). The effect of the lossy-differentiator on the sliding mode operation of the class-D amplifiers BMA and TMA is illustrated in Fig. 6.19. As it can be appreciated, the pulse-width modulated signal (PWM) still toggles when the switching function $s(V_{e1}, t)$ reaches the hysteresis voltage κ, but it exceeds the hysteresis boundary until it equates the voltage V_A. This effect is due to the lossy-differentiator pole which creates an exponential-shaped waveform instead of the triangular shape of the ideal switching function as shown in Fig. 6.18.

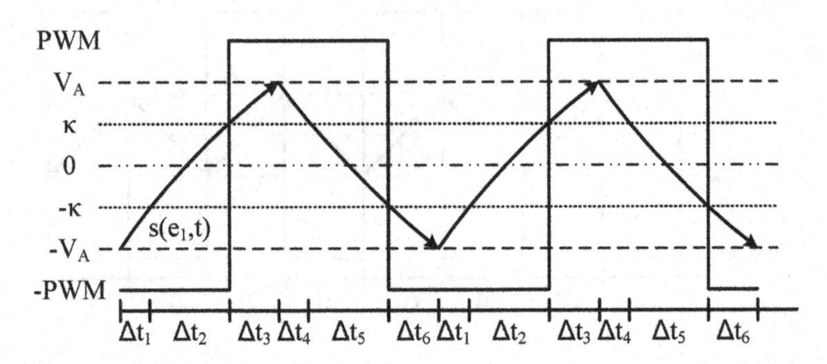

Fig. 6.19 Magnified view of real sliding mode operation for class-D amplifiers.

The value of the voltage V_A increases when the pole ω_p decreases, i.e. the switching function is very lossy, and it tends to the hysteresis voltage κ when the switching function approaches to the ideal case. As a consequence of this, the time that takes to the switching function to recover and change direction is longer and consequently the switching frequency is lower when a lossy-differentiator is employed.

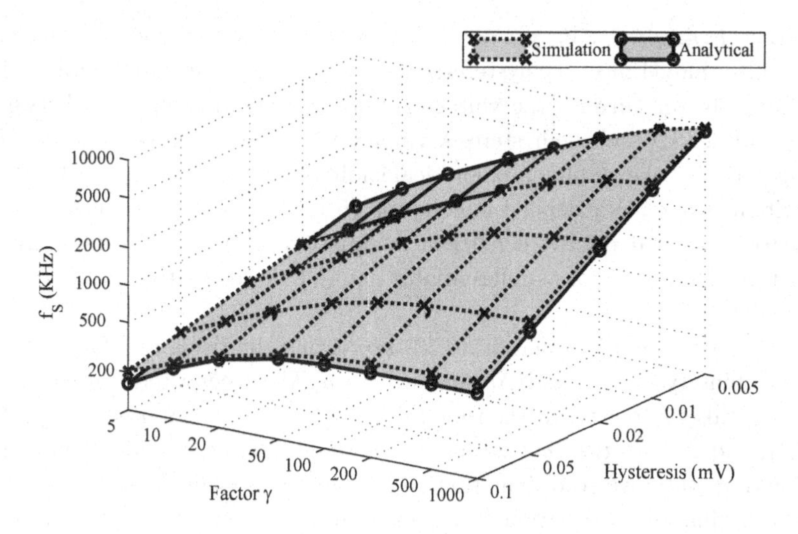

Fig. 6.20 Class-D amplifier switching frequency versus lossy-differentiator factor γ and hysteresis-window width.

A complete cycle of the lossy sliding mode operation, as shown in Fig. 6.19, can now be divided into six different subintervals of operation. Subintervals Δt_1, Δt_3, Δt_4, and Δt_6 occur when the absolute value of $s(V_{e1}, t)$ is higher than the hysteresis voltage κ. These subintervals are dominated by an exponential behavior. On the other hand, subintervals Δt_2 and Δt_5 take place when $| s(V_{e1}, t) |$ is smaller than the hysteresis voltage κ. They resemble the two subintervals of operation in Fig. 6.18 because the slope of $s(V_{e1}, t)$ within those subintervals can be considered constant. Therefore, the switching period, i.e. the inverse of the switching frequency, of the proposed class-D audio power amplifiers, derived in Appendix C, can be expressed as

$$T_{s,real} \approx \frac{2\kappa V_{DD}\frac{R}{L}\left(1 + \frac{1}{\gamma}\right)}{\left(\frac{R}{L}v_C + \frac{1}{2\gamma C}i_L\right)\left(V_{DD}\frac{R}{L}\left(1 + \frac{1}{\gamma}\right) - \left(\frac{R}{L}v_C + \frac{1}{2\gamma C}i_L\right)\right)} - 4\frac{\alpha}{\gamma}\ln\left(\frac{V_H - \kappa}{\gamma V_{e1}}\right) \tag{6.13}$$

where

$$V_H = V_{e1}\gamma \exp(-k_t) + \kappa[1 - \exp(-k_t)], \tag{6.14}$$

$$k_t = -\frac{\gamma}{4\alpha \ln(0.01)}\left(\frac{1}{f_{s,ideal}}\right) \tag{6.15}$$

and $R, L, \kappa, v_C, i_L, V_{DD}, V_H, f_{s,ideal}$, and V_{e1}, are the loudspeaker load, the filter inductance, the hysteresis window, the filter capacitor voltage, the filter inductor current, the voltage supply, the voltage difference between the voltage V_A and the hysteresis window κ, the minimum possible switching period of the amplifier under ideal sliding mode operation, i.e. the ideal switching frequency defined in equation (6.12), and the error voltage. The factor α is the derivative coefficient in equation (6.3), and γ is the product of the pole in the lossy-differentiator ω_p and the derivative factor α, i.e. $\gamma = \alpha\omega_p$.

The first term in equation (6.13) represents the rising and falling time for subintervals Δt_2, and Δt_5 in Fig. 6.19. The second term in equation (6.13) takes into account the time taken by the four subintervals Δt_1, Δt_3, Δt_4, and Δt_6. Notice that when γ tends to infinite, i.e. the ideal switching function, the inverse of equation (6.13) simply becomes equation (6.12). The evaluation of equation (6.13) for different values of γ and hysteresis voltages is plotted in Fig. 6.20 along with simulated results. Observe that the analytical prediction matches very well the simulation data. Also, notice that the switching frequency increases when γ increases and the hysteresis voltage decreases. A transversal view of the same plot is shown in Fig. 6.21.

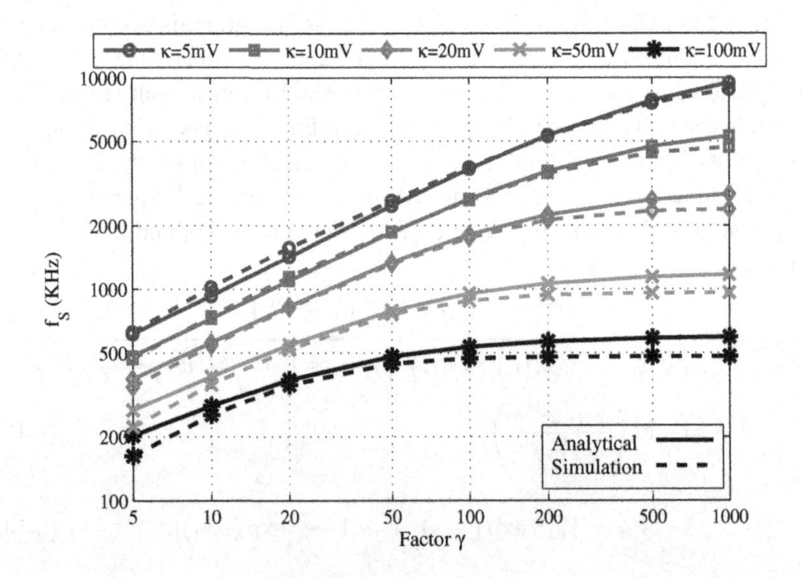

Fig. 6.21 Transversal view of class-D amplifier switching frequency versus lossy-differentiator factor γ and hysteresis-window width.

6.2.4 *Design of building blocks*

The blocks marked as I, II, III and IV in the BMA in Fig. 6.13, and the TMA in Fig. 6.15, are the sliding mode controller with feedback factor β, the comparator, the output power stage, and the low-pass filter, respectively. Both amplifiers are implemented with

$$\alpha = R_A \times C_C \tag{6.16}$$

where $R_A = 300\ k\Omega$, and $C_C = 18.75\ pF$, based on a Bessel approximation [78]. The lossy-differentiator has $R_C = 0.1 \times R_A$ to effect $\omega_p = 1/R_C C_C = 10/\alpha$ in equation (6.4). The factor $(1 + \beta)$ is given by

$$(1+\beta) = \frac{R_A}{R_B} \tag{6.17}$$

We choose $\beta = 0.4 (R_A/R_B = 1.4)$ for a reasonable compromise between linearity and output power. A simple design flow is listed in Table 6.1.

Table 6.1 Simple design flow given β, ω_p, and α.

1. Choose C_C
2. $R_A = \alpha \ / \ C_C$
3. $R_B = R_A \ / \ (1 + \beta)$
4. $R_C = 1 \ / \ \omega_p C_C$

The class-D audio amplifiers requires the implementation of a fully-differential operational amplifier in the BMA, a single-ended operational amplifier in the TMA, comparators and an output power stage. Both operational amplifiers are based on a two-stage structure with Miller compensation scheme.

The frequency response and power consumption characteristics of both, fully-differential and single-ended, operational amplifiers are specified in Table 6.2.

Comparators in both BMA and TMA amplifiers use internal positive feedback [102], and their hysteresis window is set to make the class-D amplifiers to switch approximately at 500 kHz. The comparators specifications, voltage hysteresis window, and power consumption, are listed in Table 6.3. Additionally, the output power stage is optimized [74] in order to maximize amplifier efficiency. The transistors size, tapering factor (T), and number

Table 6.2 Specifications of operational amplifiers in BMA and TMA architectures.

Parameter	OPAMP (BMA)	OPAMP (TMA)
DC gain	67 dB	74 dB
GBW	29 MHz	26 MHz
Phase margin	75°	71°
I_Q	171 μA	194 μA
P_Q	462 μW	523 μW

of stages (N), are calculated considering the short-circuit current during transitions, switch on-resistance, and parasitic capacitances.

Table 6.3 Specifications of comparators in BMA and TMA architectures.

Parameter	Comparator (BMA)	Comparator (TMA)
Hysteresis voltage	10 mV	18 mV
I_Q	80 μA	21 μA
P_Q	216 μW	57 μW

Table 6.4 summarizes the parameters of the output power stage, where W_P and L_P are the width and length of the last PMOS transistor in the buffer, respectively. Since the mobility ratio between PMOS and NMOS transistors is approximately three, it is possible to calculate the size of all the remaining transistors in the power stage from the data in the table.

Table 6.4 Characteristics of the output power stage in BMA and TMA architectures.

Parameter	Value
W_P	34560 μm
L_P	0.6 μm
T	14
N	4

The resulting design values are $\alpha \approx 5.625 \cdot 10^{-6}$, $\omega_p \approx 1.75 \cdot 10^6$ rad/s, $\omega_{3dB} \approx 125 \cdot 10^3$ rad/s, and $\beta = 0.4$.

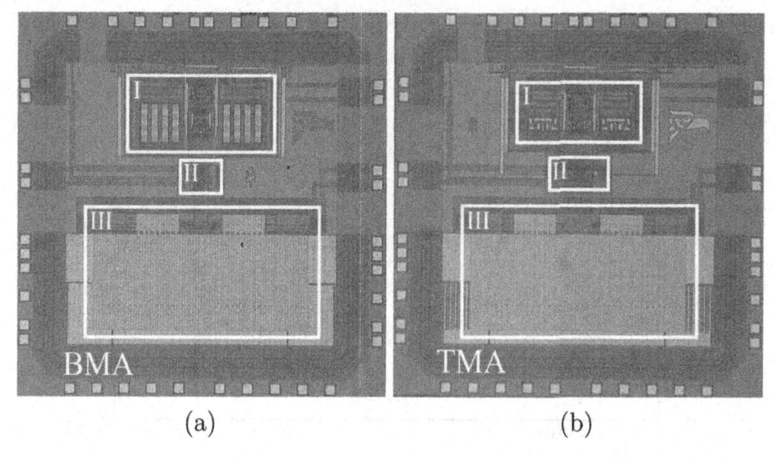

Fig. 6.22 Die micrographs (a) Binary modulation amplifier (BMA) and (b) Ternary modulation amplifier (TMA).

6.2.5 *Experimental results of CDA with SMC*

The BMA and TMA were fabricated in standard 0.5 μm CMOS technology, and the circuits were tested with the System One Dual Domain Audio Precision equipment, using a 2.7 V voltage supply. Figure 6.22 shows the BMA and the TMA die micrographs where blocks I, II and III represent the sliding mode controller SMC, the comparator, and the power stage, respectively. The total area occupied by the class-D audio amplifiers is approximately 1.49 mm^2 for the BMA, and 1.31 mm^2 for the TMA.

The efficiency (η) performance of the class-D audio power amplifiers, obtained with a sine wave input signal at 1 kHz, is shown in Fig. 6.23. The efficiency behavior of both amplifiers is comparable since the output stages are similar in both architectures. Figure 6.23 also illustrates the linearity of the amplifiers with a 1 kHz input signal. Notice that the TMA performance degrades at high output power due to its single-ended architecture. As shown in Fig. 6.24, measured power-supply rejection ratio (PSRR) is above 75 dB at 217 Hz with a sine wave ripple of amplitude 100 mV on the power supply. Signal-to-noise ratio (SNR) greater than 90 dB was measured for both class-D amplifiers.

Figure 6.25 displays the output spectra of both class-D audio power amplifiers with a 1 kHz 500 mV input voltage. Note that the harmonic components of the TMA are smaller than the BMA, but the noise floor is lower in the latter case, as expected from Fig. 6.24, because the signal-to-noise

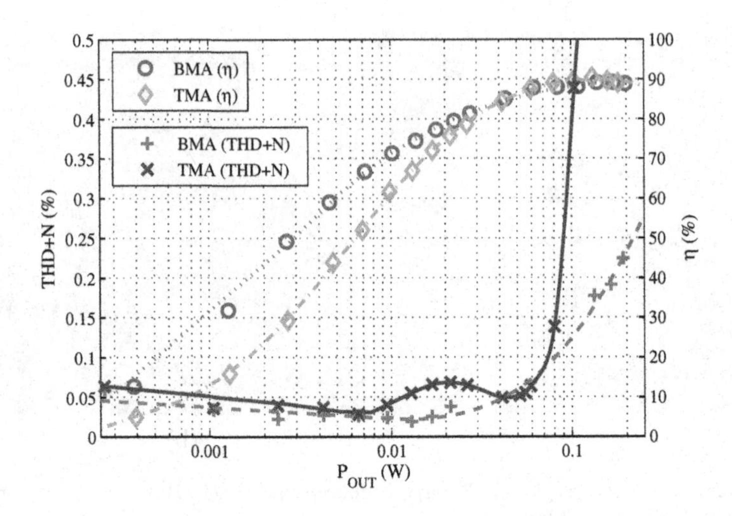

Fig. 6.23 Class-D audio power amplifiers efficiency/THD+N versus output power.

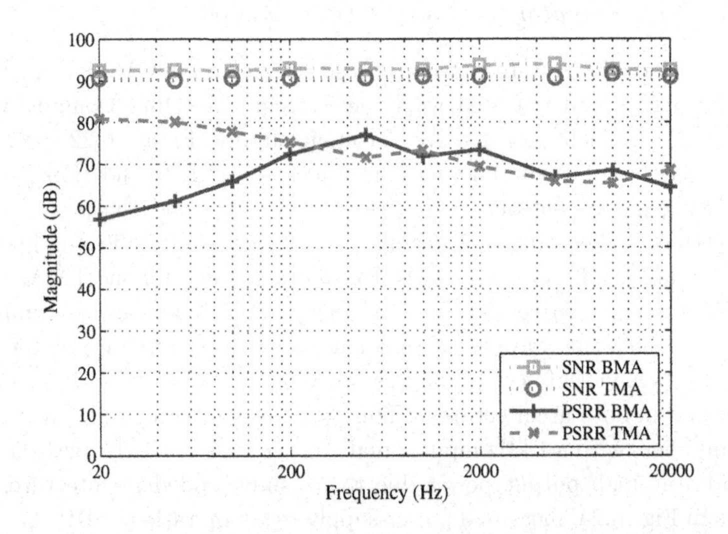

Fig. 6.24 CDAs SNR/PSRR versus frequency.

ratio (SNR) performance of the BMA is better than the TMA.

Table 6.5 compares the performance of the proposed BMA and TMA class-D audio power amplifiers to that of the state-of-the-art amplifiers.

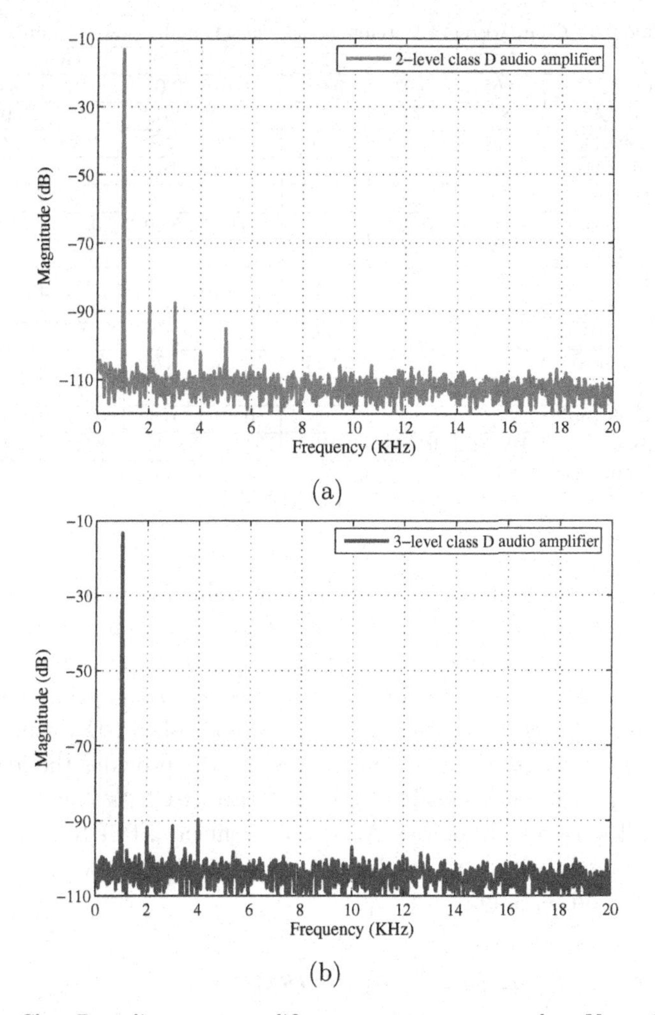

Fig. 6.25 Class-D audio power amplifiers output spectrums when $V_i = 1\ V_{pk-pk}$ (a) BMA and (b) TMA.

6.3 Integral sliding-mode control for class-D amplifiers

Architectures using variable structure control (VSC) based on sliding mode control (SMC) can decrease the power consumption, achieve low distortion, and reduce the complexity of the system [78,80,82]. Still, this approach is prone to high-frequency noise, as it requires a differentiator in the feedback loop, as discussed previously. Also, this topology has a limited power supply rejection ratio (PSRR) in the audio band because the differentiation's

Table 6.5 Comparison of state-of-the-art class-D audio power amplifiers.

Design	[57]§	[56]	[123]	[77]†	[80]	BMA	TMA
THD(%)	0.65	0.03	0.04	0.001	0.08	0.02	0.03
η(%)	85	76	79	88	91	89	90
Supply(V)	5.0	4.2	3.6	5.0	2.7	2.7	2.7
Load(Ω)	8	8	8	6	8	8	8
I_Q(mA)	16	4.7	2.5	10	2	0.25	0.21
P_Q(mW)	80	14.98	9	50	5.40	0.68	0.58
SNR(dB)	83	98	–	–	65	94	92
PSRR(dB)	60	70	84	67	70	77	81
f_s(MHz)	–	0.41	0.25	0.45	0.5	0.45	0.45
P_{OUT}(W)	1.2	0.7	0.5	10	0.2	0.25	0.25
Area(mm^2)	–	0.44	2.25	10.15	4.70	1.49	1.31
Levels	2	3	3	2	2	2	3
CMOS(μm)	–	0.9	1.2	0.6	0.5	0.5	0.5
Topology	PWM	PWM	PWM	$\Delta\Sigma$	SMC	SMC	SMC

§Commercial product.
†12 V PVDD, 10 W design, special BCDMOS process.

low-frequency attenuation reduces the loop gain. To overcome this limitation, we present a CDA with integral sliding mode control (ISMC) [92] to increase the low-frequency loop gain above that in previous architectures [78, 80, 82] and to keep the controller power consumption low.

This section presents a clock-free current-controlled CDA using integral sliding mode control [79, 81]. The proposed CDA provides the low distortion and high efficiency benefits of state-of-the-art CDAs, but consumes at least 30% less controller power. Also, improvement of PSRR is obtained by higher loop gain within the audio band when compared with [78, 80, 82], as well as good matching techniques.

6.3.1 *Class-D architecture with ISMC*

Figure 6.26 shows the block diagram of the proposed architecture. This topology consists of two feedback loops and four main building blocks. The outer voltage loop minimizes the voltage error between the input and output audio signals, and the inner current loop contains information proportional to the inductor current which is necessary to implement the controller, as will be explained later. The building blocks are the integral sliding mode controller, a hysteretic comparator, an output stage, and an off-chip low-pass filter (LPF). The ISMC processes the necessary information to generate the binary modulated signal. The hysteretic comparator obviates the carrier signal generator that would have been required in conventional architectures based on PWM [56]. The output stage provides the required current-drive capability for an 8-Ω loudspeaker, and the output filter recovers the audio signal.

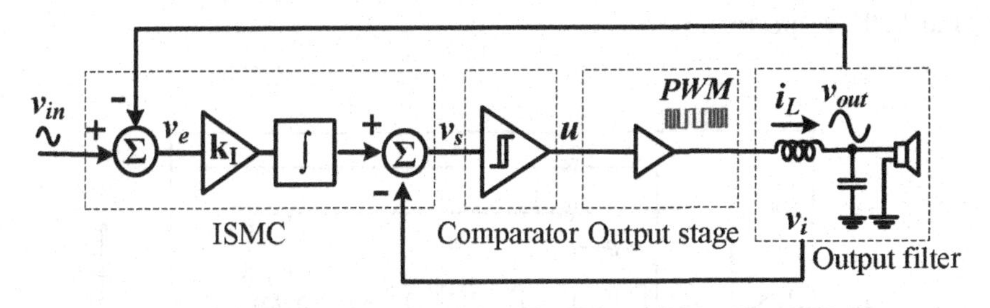

Fig. 6.26 Block diagram of CDA with ISMC.

The audio amplifier implements a tracking system governed by a control law, as shown in Appendix B, defined with the switching function given by,

$$s(V_e, V_I) = k_I \int V_e(t)dt - V_I(t) \tag{6.18}$$

where k_I is an integration constant whose value ensures stability and fast transient response, $V_e(\mathrm{t})$ is the voltage error function defined as,

$$V_e(t) = V_I(t) - V_O(t) \tag{6.19}$$

and $V_I(t)$ is a sensed voltage proportional to the inductor current $i_L(t)$.

The ISMC retains all the properties of variable structure control (VSC) with sliding-mode operation such as simple design, stability, robustness, and good transient response. Moreover, the ISMC forces the system to operate with sliding mode under any initial condition [92]. This property guarantees robust system operation from any starting point. The ISMC's integrator nulls the steady-state voltage error, and the closed-loop dynamics reduce high-frequency noise. Furthermore, sensing the current across the output inductor improves the dynamic response of the amplifier [93].

The system can be proven to be asymptotically stable with the equivalent control method analysis [92]. This method consists of determining the dynamics of the system on the switching surface, i.e. $s(V_e, V_I) = 0$. The sliding-equilibrium point of the proposed architecture is a stable focus because the eigenvalues of the system are complex with negative real part. Moreover, the final value theorem (FVT) shows that the steady-state response of the equivalent control model tracks the input signal [78], as shown in Appendix B.

6.3.2 *Integral sliding mode controller*

Figure 6.27 shows the schematic of the implemented CDA. The blocks marked as I, II, III and IV are the ISMC, comparator, output power stage, and LPF, respectively.

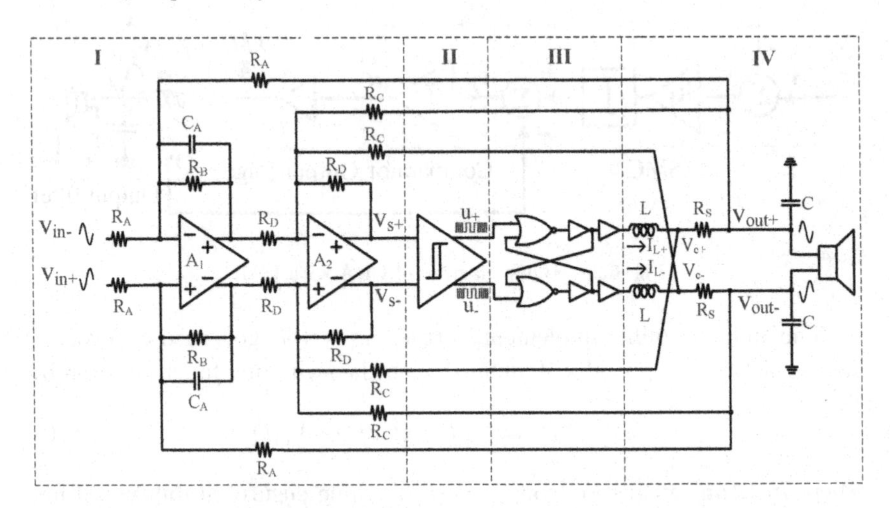

Fig. 6.27 Proposed ISMC implementation, I ISMC implementation, II comparator, III output power stage, IV LPF.

Examining the node $V_{s\pm}(t)$ one obtains the switching function implemented as

$$s(V_e, V_I) = k_I \int [V_{I\pm}(t) - V_{O\mp}(t)]\, dt - V_I(t), \tag{6.20}$$

where

$$\begin{aligned} V_I(t) &= k_s \cdot R_s \cdot i_L(t) \\ &= k_s\, [v_{c\pm}(t) - V_{O\pm}(t)] \\ &= k_s\, [v_{c\pm}(t) + V_{O\mp}(t)] \end{aligned} \tag{6.21}$$

represents the voltage proportional to the current $i_L(\mathrm{t})$ across the inductor and $k_s = R_D/R_C$. Equations (6.20) and (6.21) describe the implemented controller circuit. The CDA uses two external precision resistors (R_s) in series with the filter inductor to sense the inductor current and to feed it back to the controller. The value of these resistors was chosen high enough to sense the voltage across the resistor but sufficiently small to minimize its impact on the power efficiency of the system.

The tradeoff between the R_s value and the efficiency of the CDA when V_I = 2 V_{pp} is important; the smaller R_s, the higher the efficiency. However, an excessively small value of R_s could be comparable to parasitic board/package resistances, reducing measurement accuracy. We choose R_s = 100 mΩ to achieve both good accuracy and high efficiency, we choose k_s= 10 to have a voltage $V_I(t)$ directly proportional to i_L(t). Note that other current sensing techniques, as discussed in Chapter 4, could be employed in the ISMC architecture to improve efficiency and/or to reduce the external component count.

A fully differential amplifier (A_2) senses the inductor current using cross connected $v_{c\pm}$(t) nodes. Both the lossy integrator (A_1) and current sense (A_2) amplifiers are two-stage-Miller compensated and consume 35 μA and 90 μA of static current, respectively. Amplifier (A_1) has a DC open-loop gain of 68 dB and a phase margin of 59° and amplifier (A_2) has a DC open-loop gain of 62 dB and a phase margin of 45°.

The lossy-integrator has $k_I = 1/R_AC_A = 1.78{\cdot}10^5$ for fast transient response [78], where R_A = 280 kΩ , and C_A = 20 pF. Resistor R_B was implemented with a T-network structure to save die area. The comparator consumes only 50 μA and has internal positive feedback [124] to generate a ±10 mV hysteresis window such that the CDA runs at approximately 380 kHz.

The output buffer was designed to minimize the dynamic power dissipation without degrading the propagation delay, and we reduced the short-circuit current with a non-overlap configuration. In addition, we minimize conduction losses by reducing the CMOS on-resistance R_{on}. The calculations yielded a tapering factor between stages $T = 11$, a number of inverters $N = 4$ with and R_{on} = 220 mΩ. The dimensions of the PMOS power switch are W = 27000 μm and L = 0.6 μm and the dimensions of the NMOS power switch are W = 9000 μm and L = 0.6 μm [78–80].

The off-chip 2nd-order LPF was designed with a cutoff frequency of 20 kHz, with L = 45 μH, C = 1.5 μF, and an 8 Ω speaker. We chose a Butterworth filter approximation to achieve flat magnitude response within the audio band.

The design of the integral sliding mode controller relies on the value of the elements in the low-pass filter as mentioned in the Appendix 2. Therefore, the proposed topology could be if necessary converted into a filterless architecture by calculating the coefficients of the integral sliding mode controller according to the speaker model to obtain the highest performance possible.

6.3.3 *Experimental results of CDA with ISMC*

The class-D audio power amplifier was fabricated in 0.5 μm CMOS standard technology (V_{THN} = 0.7 V, V_{THP} = -0.9 V) and tested with a System One Dual Domain Audio Precision instrument using a 2.7-V single voltage supply. The chip was encapsulated in a DIP 40 package. Figure 6.28 shows the die micrograph of the fabricated CDA where blocks I, II, and III correspond to the ISMC, comparator, and output power stage, respectively. The total active area occupied by the class-D audio amplifier is approximately 1.65 mm^2.

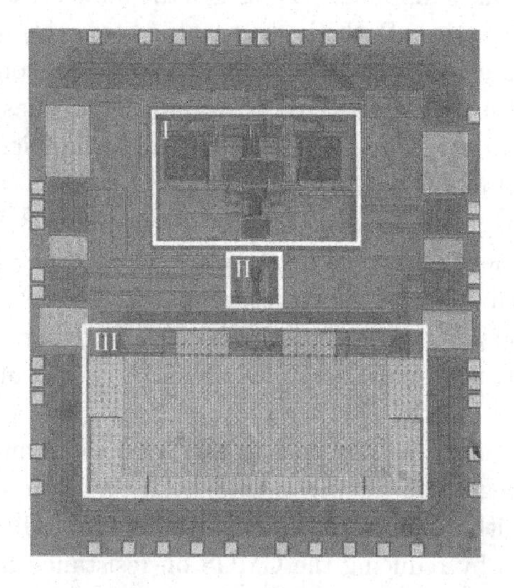

Fig. 6.28 Class-D with ISMC die micrograph, I controller (0.430mm^2), II comparator (0.033mm^2), and III output stage (1.190mm^2).

The output spectrum of the system with V_I = 2.82 V_{pp} at 1 kHz is illustrated in Fig. 6.29. As shown in the figure, the difference between the fundamental tone and the higher harmonic ($HD_3 = 3f_{in}$) is > 70 dB.

The total harmonic distortion plus noise (THD+N) and the efficiency (η) performance of the CDA are shown in Fig. 6.30 and Fig. 6.31, respectively. A THD+N of 0.02 % and an efficiency of 84 % were measured.

The proposed system achieves a maximum output power of 410 mW for 7 % THD+N. Thus, the system can provide approximately 90 % of the maximum theoretical power.

The voltage drop across R$_s$ limits the maximum output voltage swing and hence limits the maximum power. Figure 6.32 shows the PSRR and

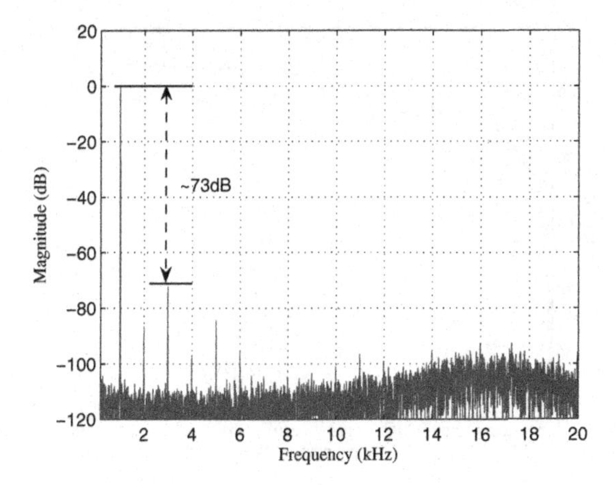

Fig. 6.29 Class-D audio amplifier output FFT when $V_I = 2.82\ V_{pp}$ at 1 kHz.

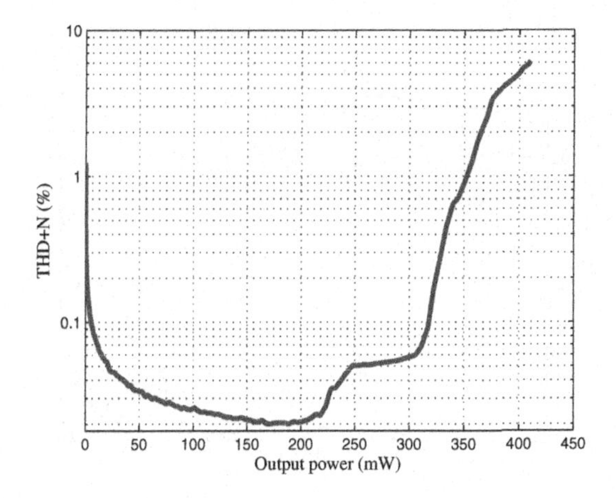

Fig. 6.30 Class-D amplifier with ISMC THD+N versus output power.

SNR versus frequency. A maximum PSRR of 82 dB was obtained while applying a sine-wave ripple of 100 mV_{pp} on the power supply. The SNR was measured with respect to 410 mW into an 8 Ω resistor and was better than 90 dB across the entire audio band.

Class-D audio amplifiers may experience power-supply-induced intermodulation distortion (PS-IMD). The PS-IMD test was performed with an

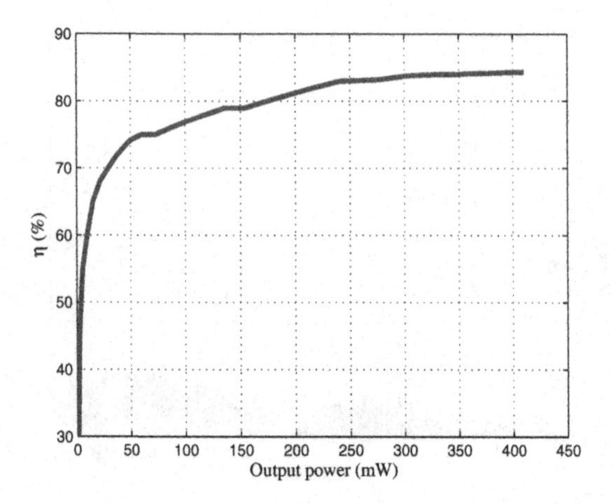

Fig. 6.31 Class-D amplifier with ISMC efficiency versus output power.

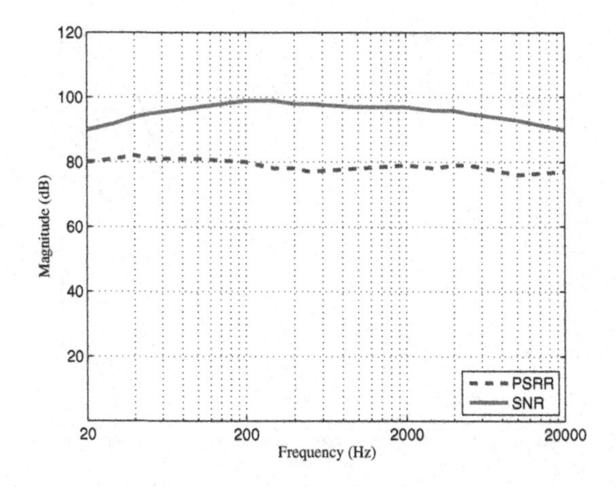

Fig. 6.32 Class-D audio amplifier PSRR and SNR versus frequency.

input voltage signal of 2 V_{pp} at 1 kHz and sinusoidal power-supply ripple of 300 mV_{pp} at 217 Hz superimposed on the DC level. Figure 6.33 shows that the difference between the intermodulation products (783 Hz and 1217 Hz) and the fundamental is approximately -90dBc.

Table 6.6 compares the performance of the CDA with ISMC to that of the state-of-the-art audio amplifiers. The presented CDA improves PSRR

Table 6.6 Performance summary for CDA with ISMC.

Design	[56]	[77]	[63]	[83]	[78]	ISMC
P_c (mW)	–	50.00	–	40.00	0.68	**0.47**
P_Q (mW)	14.98	194.00	35.00	–	–	**1.49**
I_c (mA)	–	10.00	–	8.00	0.25	**0.17**
I_Q (mA)	4.70	12.00	7.00	–	–	**0.55**
PSRR (dB)	70	67	68	70	77	**82**
SNR (dB)	98	–	102	117	94	100
THD (%)	0.030	0.001	0.01	0.001	0.020	0.02
η (%)	76	88	85	85	89	84
Supply (V)	4.2	5.0	5.0	5.0	2.7	2.7
Load (Ω)	8	6	8	8	8	8
f_s (kHz)	410	450	1800	600	450	380
P_{OUT} (mW)	700	10000	1400	1400	250	410
Area (mm^2)	0.44	10.15	–	6.00	1.49	1.65
Process	90 nm DCMOS	0.6 μm BCDMOS	–	0.7 μm CMOS	0.5 μm CMOS	0.5 μm CMOS
Topology	PWM	$\Sigma\Delta$	$\Sigma\Delta$	Hysteretic	SMC	ISMC

by at least 5 dB, and consumes at least 30 % less controller power.

6.4 Final remarks

This chapter has presented the architecture, design, implementation, and measurement of three low-power class-D audio amplifiers with a hysteretic

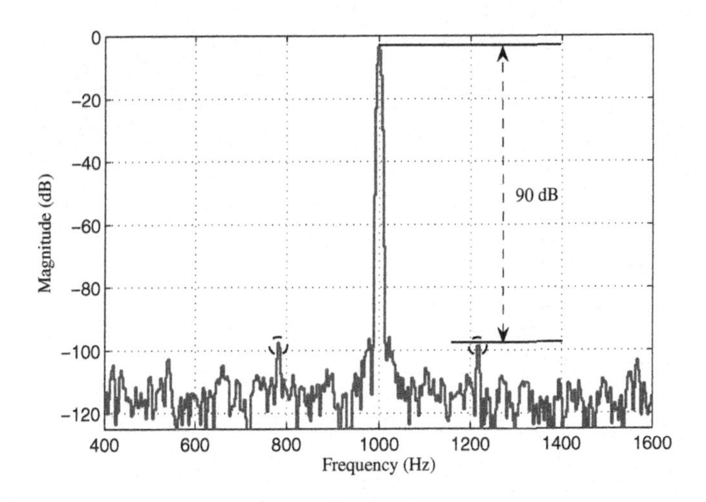

Fig. 6.33 Power supply induced intermodulation distortion measurement.

non-linear control. The prototypes have linearity, efficiency, signal-to-noise ratio (SNR), and power-supply rejection ratio (PSRR) performance comparable to the state-of-the-art works but consuming an order less of static power.

Sliding mode controllers have been implemented using lossy differentiators to limit high frequency noise amplification. Also, integral sliding mode has been used as an alternative to replace differentiators with integrators in the controller. They provide higher loop gain and limit in-band noise. However, accurate current sensing is needed for high performance applications. The amplifiers were fabricated in 0.5 μm CMOS standard technology.

Chapter 7

Class-D Output Stage for Piezoelectric Speakers

7.1 Motivation for piezoelectric speakers

The consumer's demand for smartphones and tablet computers with longer battery life has required manufacturers to implement the standard multimedia tasks, such as audio reproduction, using high-efficiency circuits. Switching DC-DC converters have been used in power management modules to achieve high-efficiency power conversion in battery-powered devices [16, 42, 87, 125]. The CDA uses a similar switching output stage as DC-DC converters to provide outstanding audio performance with high efficiency; but, to truly extend battery life, low power consumption is also required when the system is active. Conventional electromagnetic (EM) loudspeakers used in mobile devices require large amounts of power to operate, thereby limiting the battery life despite the amplifier's high efficiency.

The preferred loudspeaker for portable applications is the EM speaker. However, as discussed in Chapter 1, its electrical impedance across the audio frequency bandwidth behave as a low value resistor between 4 and 32 Ω, needing large electrical power to generate high sound pressure level (SPL). On the other hand, the piezoelectric (PZ) speaker is an electromechanical transducer that consumes little electrical power while providing high SPL in small-form factors [15]; these properties make the PZ speaker an attractive alternative to extend battery life in portable devices, especially when a high-efficiency switching amplifier such as the CDA is used [29, 126].

Closed-loop CDA architectures have been proposed to achieve high efficiency and good audio performance using different modulation techniques such as pulse-width modulation (PWM) [47, 48, 73], pulse-frequency modulation (PFM) [45, 77], or sliding-mode control (SMC) [78–82], as discussed through this book. However, these architectures have an output stage that is typically optimized to drive low impedance loads such as the EM speaker and might not be suited to provide the high-voltage output swing needed for the PZ speaker. Typical voltage levels across the PZ speaker terminals should be in the range of 10-20 V_{pp} to achieve the maximum SPL, and could be generated from the battery using high-efficiency step-up voltage circuits [16–18].

High-voltage semiconductor devices such as DMOS, LDMOS, or drain-extended MOS transistors are typically used to withstand the large voltage potential needed at the output stage. Unfortunately, these devices are typically optimized to minimize conduction losses, and their parasitic capacitors can be large, increasing the power consumption due to their large switching losses, especially when a high-frequency carrier signal is used [28,49–52]. Furthermore, using these devices in monolithic implementations would require additional fabrication steps and/or a larger silicon area; thus, increasing the cost of the amplifier. Commercial CDA architectures for PZ speakers provide high-voltage outputs using these devices, but their distortion and power consumption are still large [19–22].

Other switching output stages have been proposed to drive high-voltage capacitive actuators for different applications [30, 127]. Nonetheless, the primary objective of these applications is to deliver the maximum amount of energy at the actuator's resonant point, making them not suitable for audio applications.

This chapter discusses the design tradeoffs of the CDA for driving PZ speakers, especially when low power consumption and high efficiency are desired. An example implementation is proposed to achieve high-efficiency and high-linearity in the CDA architecture for PZ speakers to extend battery life in mobile devices [128]. The self-oscillating closed-loop architecture is used to obviate the need of a carrier signal generator to achieve high linearity with low power consumption. Moreover, the CDA monolithic implementation is able to provide an 18 V_{pp} output voltage swing in an 1.8 V core-voltage twin-well 11-16 Ω-cm p-type substrate CMOS technology without requiring expensive special high-voltage semiconductor devices. The use of stacked-cascode CMOS transistors at the H-bridge output stage provides low input capacitance to allow high switching frequency to improve linearity without sacrificing the high efficiency.

7.2 Class-D amplifier efficiency for piezoelectric speakers

The PZ speaker capacitive nature needs a different definition of power efficiency since ideally it does not dissipate average power, as discussed on Chapter 1. Thus, the typical definition of efficiency cannot be used. The amplifier's power efficiency for capacitive loads could be defined as [32–34],

$$\eta \cong \frac{P_{o,APP}}{P_{o,APP} + P_{loss}}, \tag{7.1}$$

$$P_{o,APP} = V_{o,RMS} \cdot I_{o,RMS} \cong \frac{V_{o,RMS}^2}{|Z_L|}, \tag{7.2}$$

$$P_{LOSS} = P_Q + P_{CL} + P_{SW} + P_{BD} + P_{FILT} \tag{7.3}$$

where the power dissipation in the CDA (P_{LOSS}) is mainly dominated by the amplifier quiescent power (P_Q), the conduction losses (P_{CL}), switching losses (P_{SW}) and body-diode losses (P_{BD}) of the output stage, and the losses due to the current ripple in the output filter together with the dielectric losses of the PZ speaker (P_{FILT}). The real power losses in the output filter can be expressed as,

$$P_{FILT} = I_{OUT,RMS}^2 \cdot |Z_F| \cos(\varphi) + C_{PZ} \cdot V_{OUT,RMS}^2 \cdot 2\pi \cdot F_{audio} \cdot DF \tag{7.4}$$

where $|Z_F| \cos(\varphi)$ is the resistive part of the output filter impedance at F_{SW}, C_{PZ} is the equivalent capacitance of the PZ speaker, F_{audio} is the output audio frequency applied to the PZ speaker, and DF is the dissipation factor of the PZ speaker. Typical dissipation factors range from 0.4% up to 1%.

It can be noticed that the output filter component selection affects the real power dissipation, and the DF of the PZ speaker could dominate the P_{FILT} for large operating frequencies and amplitudes. To achieve high efficiency, the CDA has to process the power of the highly reactive load with minimum power dissipation dominated by the power losses in the amplifier and output filter.

The main contributors of P_{SW} are the input and output capacitance of the output stage that can be large if the switches are sized to obtain small R_{dsON}. As discussed in Chapter 1, the advantage of using a PZ speaker is that its high impedance requires small current to operate, minimizing the impact of P_{CL} in the efficiency. This would allow smaller output switches to obtain the same P_{CL} but will decrease the P_{SW}, enhancing the overall efficiency. Moreover, the small output current together with a short $t_{deadtime}$

will reduce the P_{BD} contribution to the total power losses. To reduce the impact of the supply voltage variation due to the body-diode di/dt, a low inductance package can be used with several bonding wires in parallel for the supply, ground, and outputs.

Another consideration is that the high-voltage special semiconductor devices needed at the output stage to safely operate with the high-voltage swing required by the PZ speakers possess large input and output capacitance which would restrict the carrier signal frequency due to their large switching losses. Thus, the output stage design needs to consider the trade-offs between high voltage operation, power efficiency, and linearity.

The high-voltage switching output in the CDA for PZ speakers could impact the EMI radiated by the inductance of the cables and/or PCB traces connecting the CDA with the speaker. This is particularly important in mobile devices since most of the circuits are placed closely. Thus, the sensitive analog circuits could be drastically affected by the EMI. Several techniques to improve the EMI can be used to spread the energy of the high-frequency carrier signal used in PWM modulation, at the expense of additional power consumption and design complexity [47, 48]. The advantage of using the PZ speaker is its inherent filtering, as observed in Fig. 2.7, that can be leveraged to minimize the high-frequency energy at the output.

7.3 Class-D architecture for piezoelectric speakers

An implementation for a class-D output stage is devised using cascode devices to be able to operate at supply voltages higher than the technology nominal voltage with high efficiency. The advantage of using cascode devices at the output stage is that the input and output capacitance are reduced considerably since smaller thick oxide transistors could be used as switches to withstand the high-voltage output signal. Thus, the carrier signal frequency can be increased to enhance linearity with low power consumption. The proposed CDA architecture for driving PZ speakers with low power consumption and high linearity is shown in Fig. 7.1. A self-oscillating first-order loop was employed to avoid the extra power consumption of the modulation carrier generator, as discussed in Chapter 3. Unlike PWM or PFM modulations, the self-oscillating modulation provides inherent frequency spreading of the carrier signal to decrease the EMI without any extra power consumption.

A differential architecture was implemented to provide more dynamic range, lower distortion, and higher PSRR in the CDA. The amplifier A_1

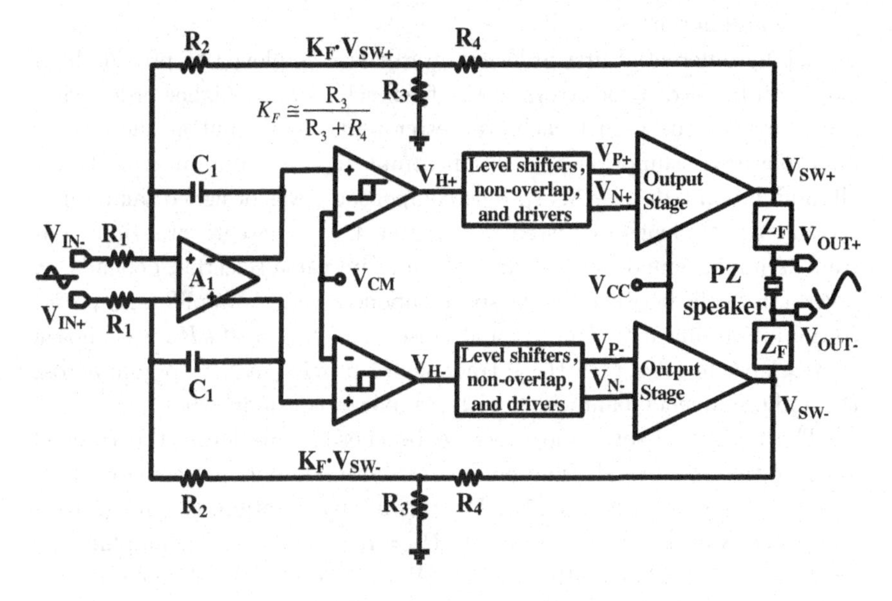

Fig. 7.1 Proposed CDA architecture for PZ speakers.

implements a first-order integrator as compensator to obtain the error signal from the difference between the input and feedback signals. The integrator's output signals are modulated by a pseudo-differential arrangement of hysteretic comparators to generate low voltage switching signals (V_{H+}, V_{H-}). These signals pass through non-overlapping, level shift, and predriver circuits, generating the gate signals for the stacked-cascode output stage. For this implementation, to achieve the desired 18 V_{PP} output signal from the H-bridge, the high-voltage supply $V_{CC} = 9\ V$ was chosen. The proposed stacked-cascode H-bridge output stage applies the high-voltage output switching signals (V_{SW+}, V_{SW-}) to the PZ speaker through an impedance (Z_F). The impedance Z_F is used in series with the PZ speaker to limit the current consumption at high frequencies and implement a low-pass filter to reduce the energy of the carrier's frequency components at the output.

Finally, the output high-voltage switching signals are fed back to the integrator using a resistive divider with factor $K_F = 1/5$ to adjust the high-voltage signal back to the nominal voltage of the technology. This selection would fix the differential closed-loop gain of the CDA to 10 V/V or 20 dB. Resistors $R_3 = 10\ k\Omega$ and $R_4 = 40\ k\Omega$ were chosen taking into account the tradeoff between their effect on the integrator's time constant and the power consumption [77].

A fully-differential first-order integrator was employed to provide high loop gain to correct for errors in the feedback loop. A higher order compensator could be used to achieve better performance but at the cost of more power consumption and design complexity to maintain stability for all modulation indexes [73, 77]. The compensator was designed taking into account the tradeoffs discussed in Chapter 4. The integrator's time constant is implemented by $\tau_I = R_1 \cdot C_1$, and its value selection depends on several tradeoffs between the passive component values, amplifier A_1 power consumption, linearity, and in-band noise. The $f_{int} = 50\ kHz$ was chosen as a compromise between these tradeoffs to provide high loop gain across the audio frequency bandwidth with low power consumption.

The input resistor values have to be chosen considering the tradeoff between the resistor's thermal noise contribution and its matching requirements for loop performance [71]. For this implementation, the integrator's component values are $C_1 = 8\ pF$ and $R_1 = R_2 = 400\ k\Omega$. The amplifier A_1 provides a DC gain of 45 dB with a GBW of 70 MHz. This design selection yields a magnitude error and phase error in the integrator function of 0.5% and 0.07% [101], respectively. The hysteresis window of the comparators in the modulator, the integrator's time constant, and the propagation delay in the loop will determine the modulation frequency of the self-oscillating system as expressed in,

$$F_{SW}(D) = \frac{D \cdot (1 - D)}{\frac{V_{hyst} \cdot \tau_{int}}{V_{supply}} + \tau_d} \tag{7.5}$$

where V_{supply} is the supply voltage of the comparator, V_{hyst} is the voltage hysteresis window of the comparator, τ_{int} is the integrator's time constant, and τ_d is the propagation delay from the comparator to the input of the integrator, including the comparator delay. The average value of F_{SW} could be chosen taking into account the tradeoffs between power consumption, distortion, output filter components, and the excitation of undesired mechanical resonant modes in the PZ speaker [15].

A higher value of F_{SW} would result in less distortion and smaller output filter components but at the expense of extra power consumption due to higher switching losses [77], and wider GBW of A_1. On the other hand, a low value of F_{SW} would reduce the power consumption but at the expense of more distortion and bigger output filter component values [70]. Leveraging the low input capacitance of the stacked-cascode output stage in the proposed class-D output stage, a high-frequency carrier is used to achieve high linearity with high efficiency. The average value of $F_{SW} = 800$ kHz was chosen as a compromise between these tradeoffs for this implementation. Therefore, from (7.5), V_{hyst} can be found for a D = 0.5 as,

$$V_{hyst} = \frac{V_{supply}}{\tau_{int}} \cdot \left(\frac{1}{4 \cdot F_{SW}} - \tau_d \right). \tag{7.6}$$

For a $V_{supply} = 1.8$ V, $F_{SW} = 800$ kHz, $\tau_{int} = 3.2$ μs, and $\tau_d = 100$ ns, a $V_{hyst} \cong 120$ mV was obtained. Figure 7.2 shows the variation of the calculated F_{SW} versus the duty cycle (D) for several τ_d cases as expressed in Eq. (3.15). It can be seen that F_{SW} is a parabolic function of D, and the delay τ_d would impose a limit in the maximum achievable F_{SW}. For the assumed $\tau_d = 100$ ns, the average switching frequency decreases from 800 kHz to 300 kHz as the peak input amplitude increases/decreases.

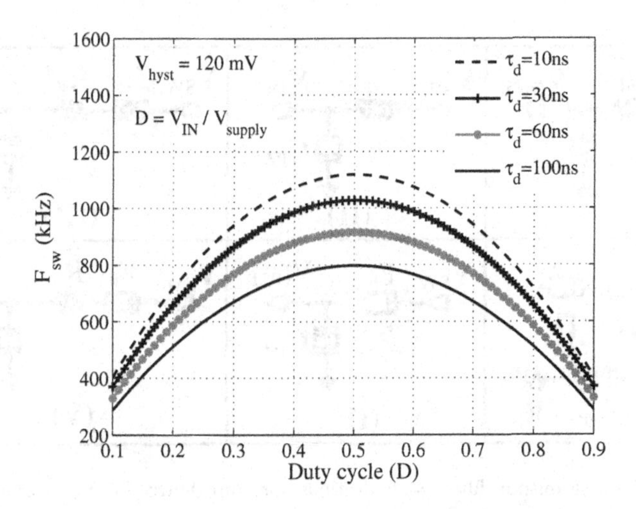

Fig. 7.2 Calculated switching frequency (F_{SW}) versus duty cycle (D) of the CDA for several propagation delay (τ_d) cases with a fix hysteresis window.

One of the drawbacks of the variable-frequency modulation is that the output current ripple will be increasing for large audio signals due to the decreasing F_{SW}. Thus, the output RMS current will increase, and the real power dissipation in the non-ideal components of the output filter will increase as expressed in Eq. (7.4). The F_{SW} variation could be reduced if needed by controlling the main parameters in Eq. (7.5) such as the propagation delay [87], the hysteresis window [88, 89], or the integration time constant [90]. For this implementation, the variable F_{SW} will be exploited to help spread the energy of the high-voltage high-frequency switching signal at the output of the audio amplifier to decrease the radiated EMI components.

Another consideration about the capacitive behavior of the PZ speaker is that it presents a low impedance value at high frequencies, especially close to F_{SW}, that will increase the current consumption of the speaker if it is driven directly by the switching output signal. To minimize this effect, an impedance Z_F can be placed in series with the PZ speaker to limit the current delivered to it at high frequencies. An additional benefit of using Z_F in series with the PZ speaker is the inherent filtering function since the PZ speaker behaves as a capacitor. This inherent output filter will mitigate the high frequency components of the high voltage switching output that could negatively impact the EMI.

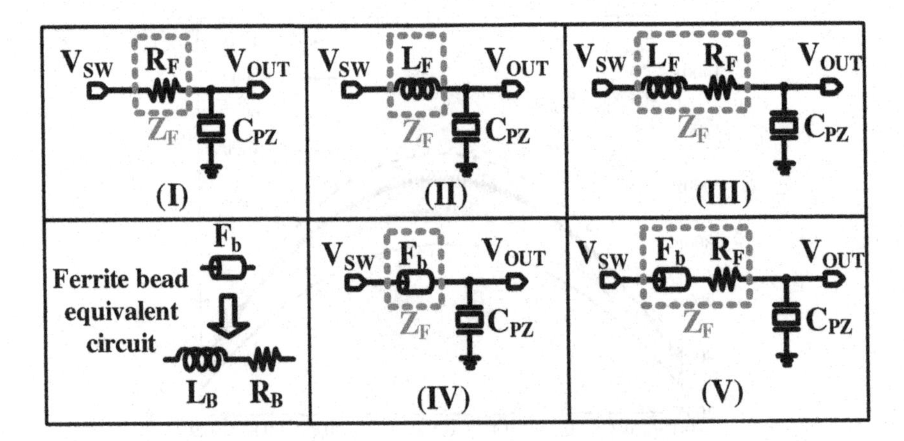

Fig. 7.3 Different output filter configurations for impedance Z_F together with PZ speaker equivalent impedance C_{PZ}.

Several options for implementing Z_F can be selected as shown in Fig. 7.3 (single-ended configurations are shown for simplicity). A current limit resistor (R_F) can be used as in Fig. 7.3(I) since its impedance is constant and independent of frequency. However, using a resistor could impact the efficiency of the overall audio system since it would dissipate power. Its value selection needs to take into account the cut-off frequency of the low pass filter, current limit, and power dissipation. The transfer function of the resulting first order RC low pass filter with cut-off frequency ω_{RC}, given by R_F and the PZ speaker impedance (C_{PZ}), is expressed as

$$\left.\frac{V_{OUT}(s)}{V_{SW}(s)}\right|_I = \frac{1}{1 + s/\omega_{RC}} = \frac{1}{1 + s \cdot C_{PZ} \cdot R_F}. \tag{7.7}$$

A more power-efficient alternative is to use a reactive element to implement Z_F. An inductor can be used as in Fig. 7.3(II). The inductor high impedance at high frequencies compensates for the low impedance of the PZ speaker to have a more constant output impedance, thereby limiting the current. Moreover, the resulting low pass filter is second order, minimizing the EMI and the carrier signal energy at the PZ speaker. R_F is used as in Fig. 7.3(III) to introduce damping in the second order filter to avoid unwanted peaking that can increase the output signal distortion.

However, the R_F value needs to be chosen taking into account its power dissipation since the inductor ripple would dissipate real power across the resistor. The transfer function of the second order low pass filter, with cut-off frequency ω_{LC} given by the L_F and C_{PZ}, is expressed as,

$$\begin{aligned}\left.\frac{V_{OUT}(s)}{V_{SW}(s)}\right|_{III,V} &= \frac{\omega_{LC}^2}{s^2 + s \cdot 2 \cdot \zeta \cdot \omega_{LC} + \omega_{LC}^2} \\ &= \frac{1/(L_F \cdot C_{PZ})}{s^2 + s \cdot (R_F/L_F) + 1/(L_F \cdot C_{PZ})}.\end{aligned} \tag{7.8}$$

Figure 7.4 shows the frequency magnitude response for different output filter configurations. It can be observed, that filter III provide the best attenuation at the EMI region, but at the expense of more external components; on the other hand, the filter I uses less external components, but at the expense of less attenuation at the EMI region and increased power dissipation.

The main drawbacks of choosing Z_F as an inductor are the component cost and PCB area occupied. On the other hand, if an application does not require a low cut-off frequency in the low-pass filter but requires low EMI, a ferrite bead (Fb) can be used as the reactive element to filter out the high frequency components as in circuit Figs. 7.3(IV) and 7.3(V). The ferrite bead behaves as an inductance (L_B) at high frequencies and as a low value resistor (R_B) at low frequencies as shown in Fig. 7.3.

Ferrite beads cost less than an inductor and use less PCB area. However, the ferrite bead selection needs to take into consideration the peak current capability of the core material to avoid current saturation and variations in the equivalent inductance that could increase signal distortion. Also, the equivalent series resistor R_B value needs to be considered to avoid extra power dissipation.

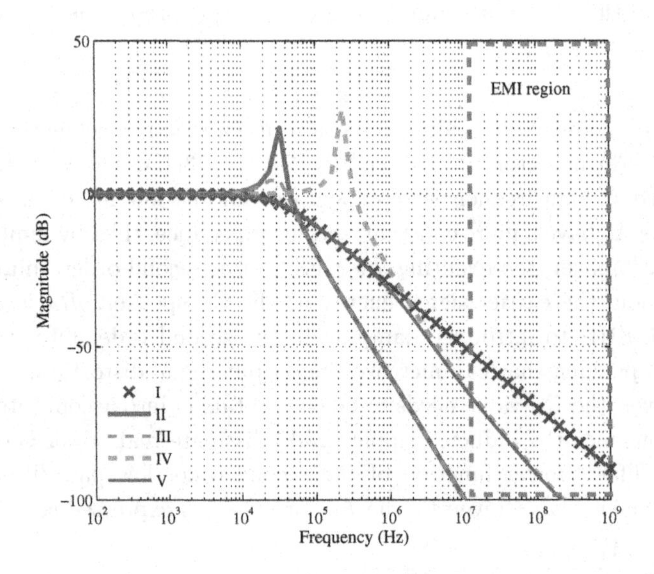

Fig. 7.4 Bode plot for output filter configurations with PZ speaker.

7.4 Stacked-cascode H-bridge output stage

The main motivation for using stacked-cascode switches in the H-bridge output stage is to reduce the switching losses due to the small input and output capacitors. Moreover, it allows to handle high voltages in monolithic implementations while ensuring sufficient lifetime in a CMOS technology with a significantly lower supply voltage. This over-voltage protection is conceptually illustrated in Fig. 7.5, where the stacked-cascode transistors absorb enough voltage across them to allow safe operation in the main switch.

The use of cascodes in the output stage of the CDA presents two main challenges. First, the switching output signal (V_{SW}) is changing between V_{CC} and GND. Therefore, two different gate voltages for the cascode transistor connected to the output terminal are required [41]. This is to avoid exceeding the maximum allowed voltage potential across any of its terminals during the output high or low state. Therefore, a simple adaptive biasing structure is proposed for this implementation to safely operate the stacked-cascode H-bridge. Second, the R_{dsON} and V_{SD} increase by adding cascodes. The impact of the conduction losses will depend on the average output current flowing through the stacked-cascode switches. Each additional stacked-cascode switch will increase the total V_{SD} that will increase the P_{BD}. However, since the PZ speakers appear as a high impedance to the amplifier, the RMS current flowing through the H-bridge is small, lessening the impact of large R_{dsON} and V_{SD} on the efficiency. The proposed

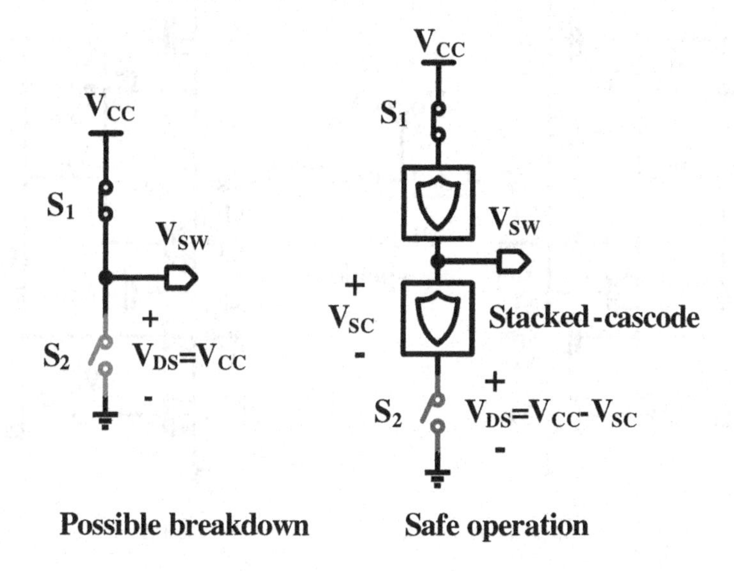

Fig. 7.5 Stacked-cascode over-voltage protection conceptual operation.

stacked-cascode output stage for driving PZ speakers is illustrated in Fig. 7.6. Thick-oxide 3.3 V transistors were used in the output stage; however to achieve safe operation the top PMOS transistor has to ensure that its N-well to substrate potential does not exceed the reverse bias breakdown voltage of 10.8 V for the CMOS technology used, limiting the V_{CC} to 9 V.

The thick-oxide devices can tolerate sustained operation within 10% of their voltage rating but they can suffer from irreversible damage if the voltage across its terminals exceed the gate oxide breakdown voltage of 5.2 V. All the transistors have their sources tied to their bulk to avoid larger R_{dsON} due to the body effect; where the NMOS cascode devices use triple-well transistors with N-well voltage (V_{NW}) of 7.5 V to reverse bias the P-well to N-well diode. The transistors M_1, M_6, M_7, and M_{12} are the input switches of the H-bridge. Two cascode transistors are stacked vertically on top of each input switch to avoid exceeding the maximum allowed voltage potential across any of their terminals. The biasing at the gates of the cascode devices ensure that when the signal switch M_1, M_{12} or

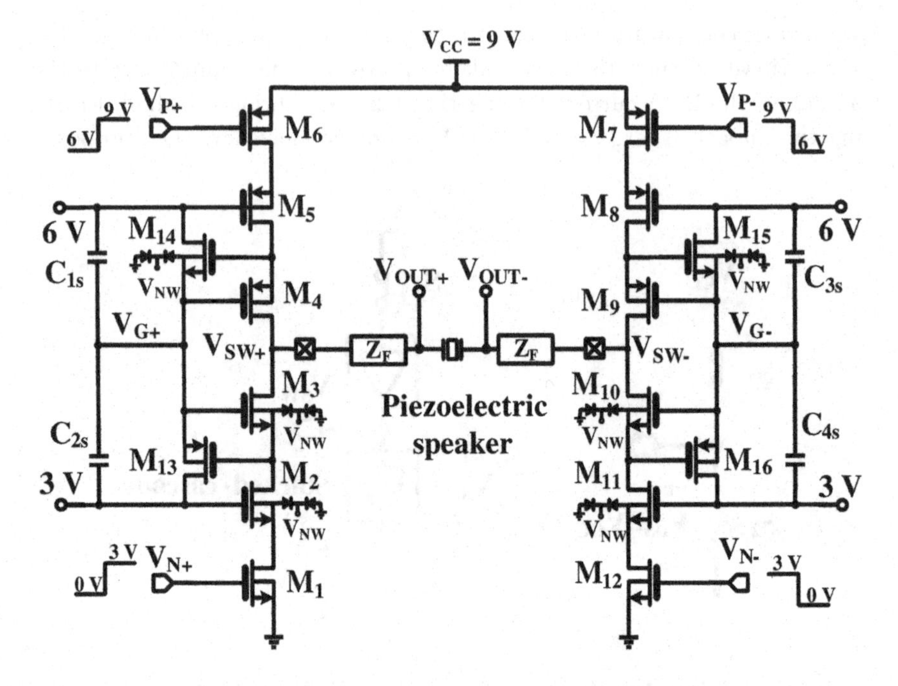

Fig. 7.6 Proposed output stage schematic for driving PZ speakers.

M_6, M_7 are off, the source voltages of the cascode devices will follow,

$$V_S > V_G - V_{TH} \tag{7.9}$$

where V_S is the source voltage, V_G is the gate voltage, and V_{TH} is the threshold voltage of the transistor.

Table 7.1 Stacked-cascode biasing design procedure.

1. Determine $V_{CC} = V_{O,RMS}/2$ for maximum SPL of the chosen PZ speaker.
2. Verify the maximum allowed rated voltage (V_{BRK}) across any terminal for the chosen devices of the CMOS technology.
3. Determine the number of devices $N = V_{CC}/V_{BRK}$ needed in series.
4. Select $V_{G,PMOS} < V_{CC} - V_{BRK} + V_{TH}$.
5. Select $V_{G,NMOS} < V_{BRK} + V_{TH}$.
6. Capacitors C_{1s} to C_{4s} are sized such that node $V_G\pm$ does not change during the high to low transition and low to high transition, respectively.

The gate voltage of M_2 and M_{11} is fixed to 3 V, while the gate voltage of M_5 and M_8 is fixed to 6 V. This is to ensure that the voltage drop across any terminal of the transistors is below 3.3 V. Table 7.1 details the design procedure for the biasing of the proposed stacked-cascode output stage for PZ speakers.

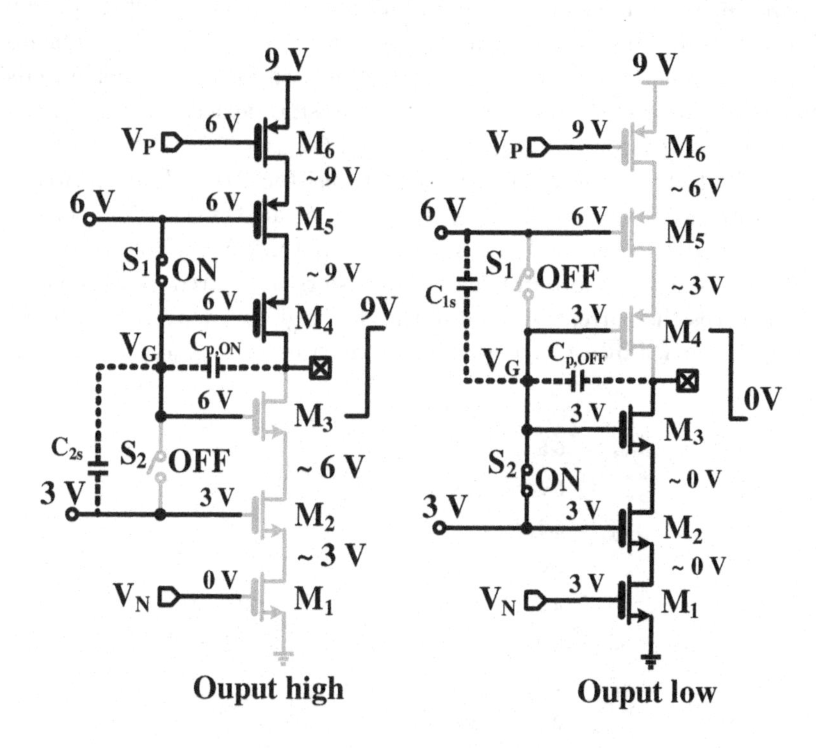

Fig. 7.7 Proposed stacked-cascode output stage simplified operation.

The output switching node V_{SW} will be switching between 0 V and 9 V; therefore, the gate of transistors M_3, M_4, M_9, M_{10} need to change its voltage to ensure the safe operation of these transistors. Transistors $M_{13} - M_{16}$ are used as switches to alternate the gate voltage (V_G) of the cascode transistors M_3, M_4, M_9, and M_{10}, between 3 V and 6 V, depending on the switching state. These gate voltages allow safe operation during the switching transient for the transistors. Capacitors $C_{1s} = C_{3s} = 9\ pF$ and $C_{2s} = C_{4s} = 2\ pF$ are used to stabilize the node V_G by absorbing the charge injected at this node during the switching transients. The steady-state operation of the stacked-cascode output stage of the two switching states for half of the H-bridge is depicted in Fig. 7.7 for the switching high state, and low state; for simplicity, transistors M_{14} and M_{13} were replaced for switches S_1 and S_2, respectively.

For the switching high state, transistors $M_4 - M_6$ and switch S_1 are ON, and transistors $M_1 - M_3$ and switch S_2 are OFF. On this operating condition, the voltage at V_G is 6 V, allowing a maximum voltage drop of 3 V across any of the terminals of transistors $M_2 - M_5$. The capacitor C_{2s} was chosen to provide a low impedance path for the current injected during the low to high transition by the parasitic capacitance $C_{p,ON}$, that is mainly composed of $C_{gd,M4}$, $C_{gs,M4}$, and $C_{gs,M14}$.

Similarly, for the switching low state, transistors $M_4 - M_6$ and switch S_1 are OFF, and transistors $M_1 - M_3$ and switch S_2 are ON. For this operating condition, the voltage at V_G is 3 V, allowing a maximum voltage drop of 3 V across any of the terminals of transistors $M_2 - M_5$. The capacitor C_{1s} absorbs the current injected during the high to low transition by the parasitic capacitance $C_{p,OFF}$ mainly comprised of $C_{gd,M3}$, $C_{gs,M3}$, and $C_{gs,M13}$.

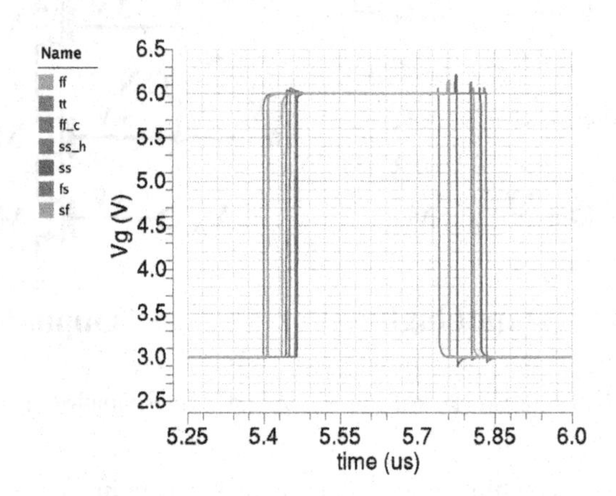

Fig. 7.8 Transient simulation of V_G across PVT.

The automatic adjustment of V_G allows a safe operation for the cascode transistors connected to the output terminal. To verify this, the circuit was simulated across process and temperature variations (PVT) for the slow PMOS and slow NMOS (ss) case, the slow PMOS and fast NMOS (sf) case, the fast PMOS and fast NMOS (ff) case, and the fast PMOS and slow NMOS (fs) case. The two extreme corners are represented by the ff case with cold temperature (ff_c) and the ss with hot temperature (ss_h). The node V_G was saved on each simulation case, and the results are plotted in Fig. 7.8. It can be observed that V_G never exceeds the 6.5 V or 2.5 V limits, avoiding stress in the transistors that could deteriorate their performance. The timing variations observed are expected since the blocks used in the implementation of the CDA will vary the loop delay τ_d by small amounts, changing the F_{SW} as shown in Fig. 7.2.

Figure 7.9 shows the block diagram for the gate driver of the stacked-cascode output transistors for half-bridge of the output stage. The goal of the gate drivers is to drive the gate capacitance of the output MOS devices.

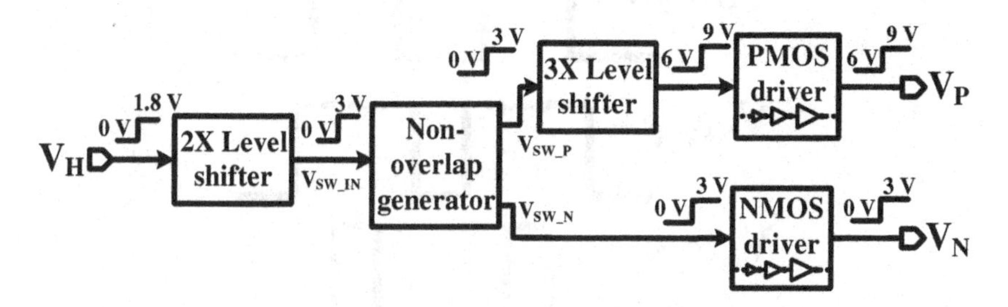

Fig. 7.9 Gate driver block diagram stacked-cascode output stage.

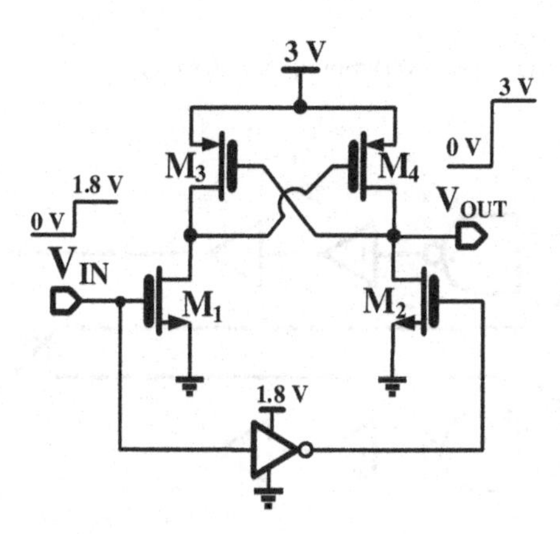

Fig. 7.10 Level shifter from 1.8 V to 3 V schematic.

The switching output of the comparators (V_H) is level shifted from 1.8 V to 3 V using a cross coupled level shifter circuit as shown in Fig. 7.10. The level shifter is implemented with 3 V thick-oxide transistors $M_3 - M_4$ and it uses positive feedback to decrease the propagation delay of the block. The output of the level shifter switches between 3 V and ground.

A non-overlap signal generator is used to avoid excessive short circuit current at the output stage. The non-overlap time of 2 ns or dead-time of 0.16% of the period is chosen as a tradeoff between the propagation delay τ_d in the loop, efficiency, and distortion [103]. The non-overlapped signals, as illustrated in Fig. 7.11, are applied to the pre-driver circuits that will drive the gates of the stacked-cascode output transistors.

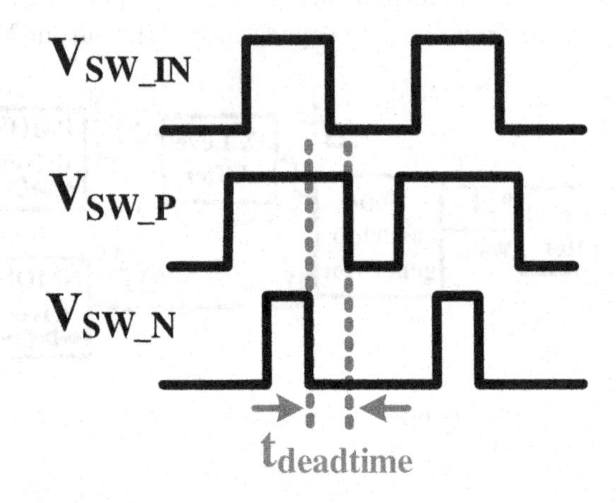

Fig. 7.11 Non-overlapping gate drive typical waveforms.

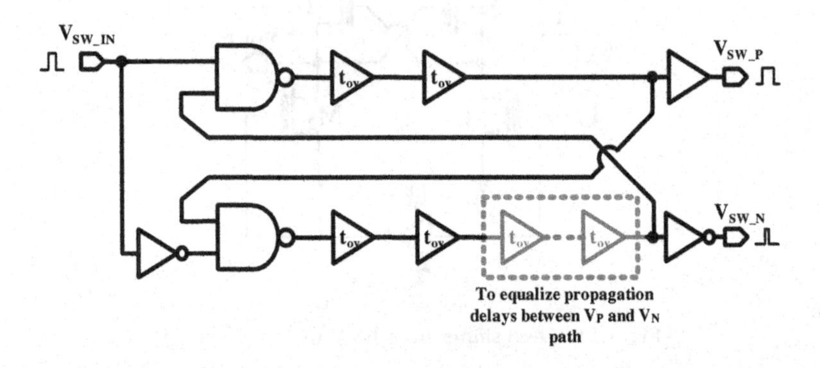

Fig. 7.12 Implemented non-overlap generator for gate drivers.

The PMOS signal path has an extra level-shift block that introduces some delay and makes the non-overlap delay at the output signals (V_P/V_N) asymmetrical. This asymmetry would cause the switching node to introduce distortion to the audio signal. To correct this, the non-overlap delay for the NMOS path was adjusted to the PMOS path by introducing extra delay elements, as observed in Fig. 7.12.

The gate signal (V_P) of the output PMOS switch connected to V_{CC} needs to switch between 6 V and 9 V to avoid excessive high voltage potential across its terminals. This is achieved by level-shifting the switching signal from 0-3 V to 6-9 V using a high-speed 3X level-shifter [129] with triple-well NMOS transistors in the inverters to shift the ground level to 6 V, as shown in Fig. 7.13. Stacked-cascode transistors are also used to protect from exceeding voltage stress limit in the main switches of the level shifter. A bootstrap capacitor C_{3X} is used to reduce the propagation delay of the block by injecting current in the positive feedback latch implemented by $M_6 - M_7$.

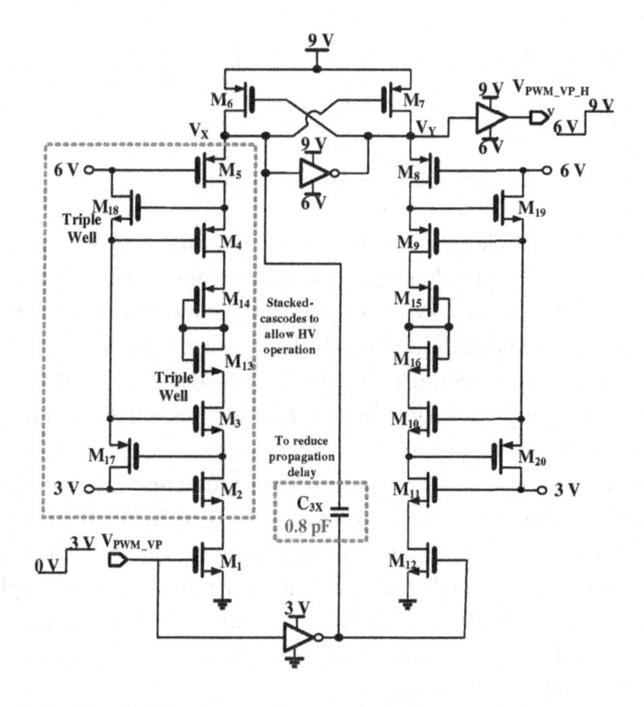

Fig. 7.13 Level shifter from 3 V to 9 V schematic with floating ground.

The PMOS drivers use floating inverters with triple well 3 V thick-oxide transistors to keep the signal switching between 6 V and 9 V, as shown in Fig. 7.14. The NMOS drivers are also implemented using 3 V thick-oxide transistors to allow the signal to switch between 3 V and 0 V. The main

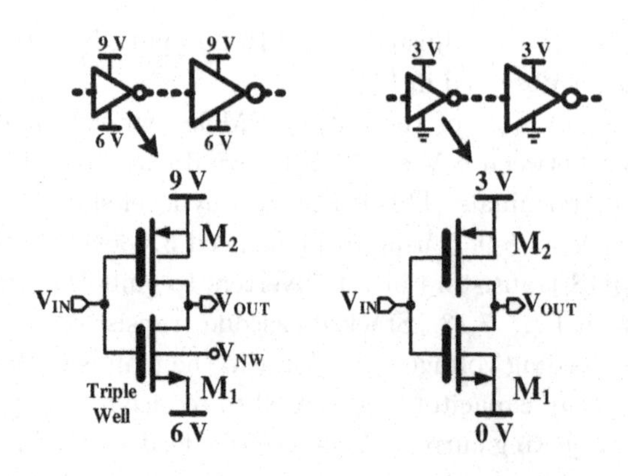

Fig. 7.14 Gate drivers implementation for PMOS and NMOS path.

goal of these gate drivers is to minimize the delay with minimum switching loss.

7.5 Experimental results for PZ speakers

The CDA for PZ speakers was fabricated in 0.18 μm CMOS standard technology [128]. Figure 7.15 shows the die micrograph of the fabricated CDA, where blocks I, II, III, and IV correspond to the integrator, hysteretic comparators, pre-driver circuits, and stacked-cascode output stage, respectively. The total active area occupied by the proposed CDA for PZ speakers is 0.4165 mm^2, where the stacked-cascode H-bridge uses 0.2571 mm^2 (61.72%) of the active silicon area.

The prototype was tested with a System One Dual-Domain Audio Precision (AP) instrument as shown in Fig. 7.16. The AP instruments provide a complete solution for characterizing audio performance. The instrument is capable to generate high audio quality output signals to apply to the device under test (DUT), and it has a signal acquisition port to capture the audio signal for processing. The setup in Fig. 7.16 allows to measure the THD+N, SNR, and output power.

To measure the supply current and efficiency of the amplifier, the input current, the output current, and the output voltage waveforms were monitored with an oscilloscope as shown in Fig. 7.17. The current waveforms were measured using series sensing resistors $R_s = 0.1\ \Omega$, where the voltage across them is proportional to the current.

The measured CDA supply current versus the output RMS voltage for

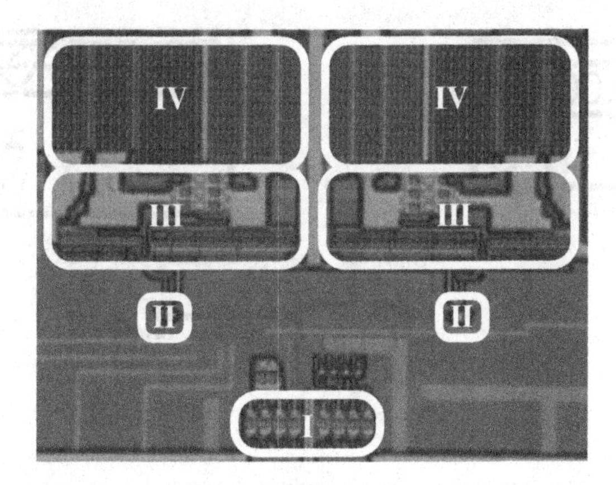

Fig. 7.15 Die micrograph of CDA for PZ speakers, I integrator (0.0715 mm^2), II comparator (0.0026 mm^2), III Pre-drivers (0.0852 mm^2), and IV stacked-cascode output stage (0.2571 mm^2).

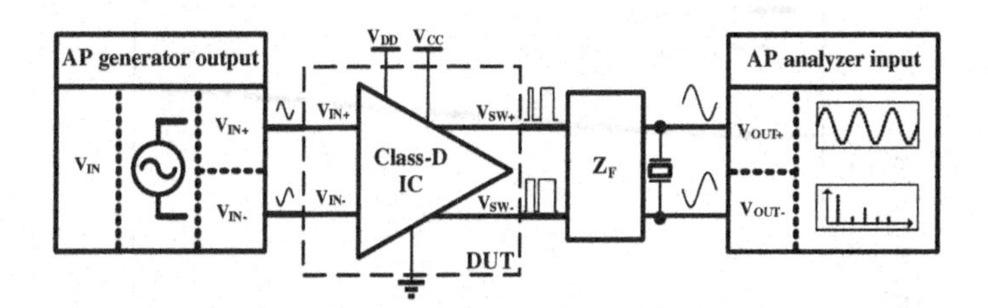

Fig. 7.16 Measurement test configuration of CDA for PZ speakers.

a 1 kHz signal is shown in Fig. 7.18. The RMS value (i) represents the capacity of the CDA to process the demanded current by the PZ speaker; the RMS current is dominated by the switching frequency ripple at lower output voltages and by the PZ speaker at higher voltages.

The average current value (ii) represents the power dissipation of the system and is expected to be very low since the load is highly reactive; it was obtained by averaging the supply current waveform for several audio signal periods. The measured quiescent supply current of the proposed CDA driving the PZ speaker at idle condition (e.g. when no audio signal is present) is 0.7 mA (average current).

The power-efficiency of the CDA with the PZ speaker was measured using the apparent power as described in Chapter 1. Figure 7.19 shows the measured efficiency versus the output RMS voltage for a 1 kHz signal, achieving a maximum efficiency of 96%.

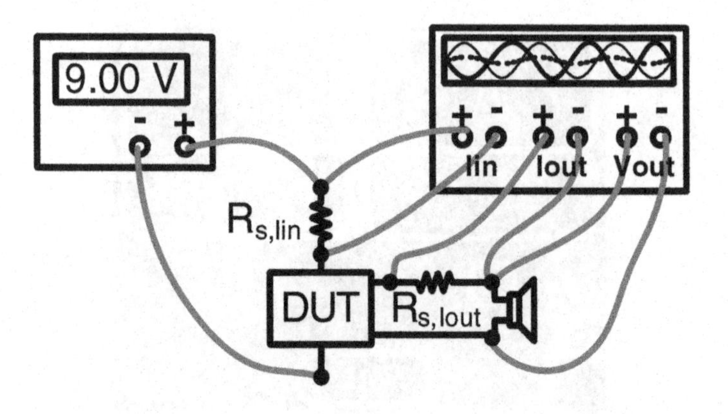

Fig. 7.17 Measurement test configuration for supply and output currents.

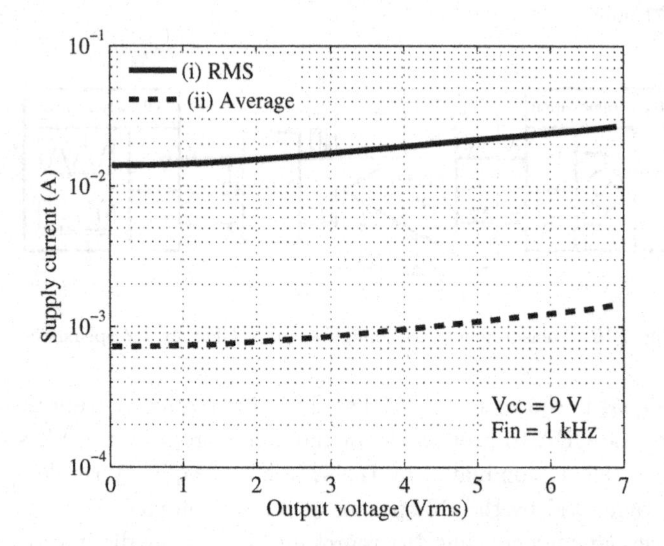

Fig. 7.18 Measured supply current for CDA driving a PZ speaker.

The measured frequency spectrum of the system for an output $V_{OUT} = 18\ V_{PP}$ at 1 kHz is illustrated in Fig. 7.20, where the difference between the fundamental tone and the highest harmonic is −67 dB. High linearity is achieved with high-voltage output swing as desired for audio applications using PZ speakers. The integrated output noise from 20 Hz–20 kHz (unweighted) for the idle condition was obtained as 167 μV.

To evaluate the impact of different reactive elements at the output filter,

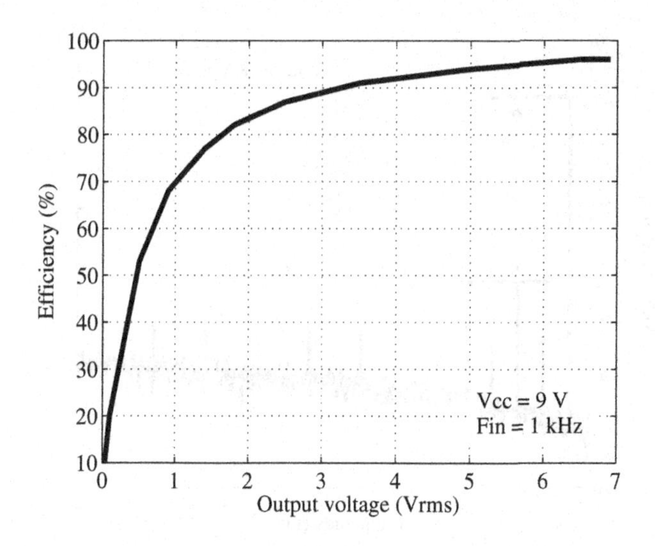

Fig. 7.19 Measured power-efficiency for CDA driving a PZ speaker.

two Z_F implementations, (III) and (V) with C_{PZ} = 470 nF, R_F = 5.6 Ω, L_F = 47 μH, R_B = 100 mΩ, and L_B = 1 μH as shown in Fig. 7.3, were used for the THD+N measurement. The measured THD+N of the proposed CDA for both Z_F configurations with a 1 kHz signal is shown in Fig. 7.21.

The system with the output filter (III) achieves better THD+N performance than the system with the output filter (V). The minimum measured THD+N is 0.025% and 0.1% for the CDA with Z_F configurations (III) and (V), respectively. The degradation in THD+N in filter (V) appears to be caused by the ferrite bead material due to its magnetic history curve (B-H curve) non-linear behavior, and a non-constant permeability (μ_m) that changes with the magnitude of the magnetic field and operating frequency [12]. The THD+N for the output filter (III) with a signal at 6.67 kHz is also included in Fig. 7.21 where a degradation in the THD+N can be observed due to the third harmonic distortion being dominant.

To measure the amplifier's power-supply intermodulation distortion (PS-IMD) performance, a 1 kHz sine wave with 0.5 V_{PP} was used as the input of the audio amplifier together with 0.2 V_{PP} at 217 Hz signal at the amplifier's high-voltage supply V_{CC}, as shown in Fig. 7.22.

The measured frequency spectrum of the output signal is shown in Fig. 7.23. As can be seen, both intermodulation tones are at least 96 dB below

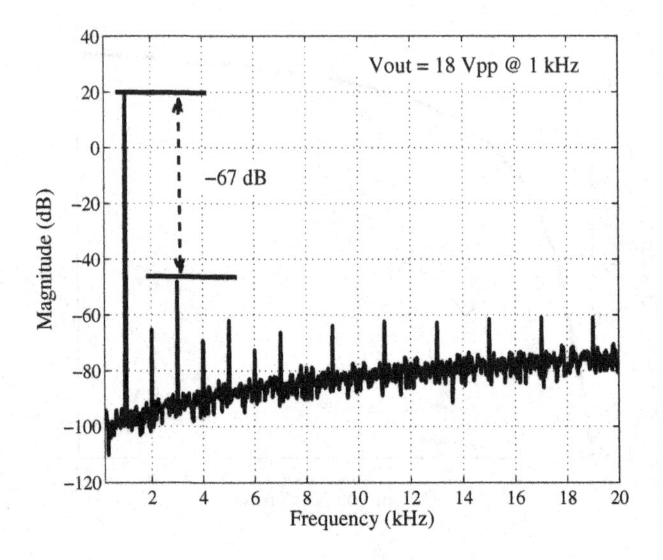

Fig. 7.20 Measured frequency spectrum for 18 V_{PP} output signal at 1 kHz.

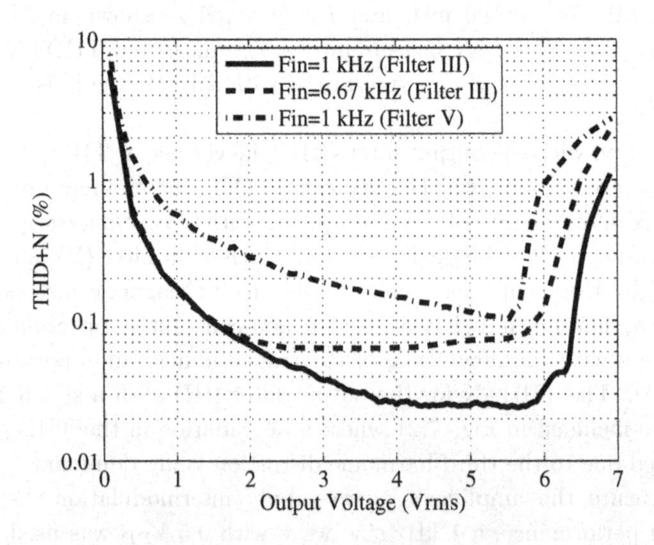

Fig. 7.21 Comparison of the measured THD+N versus output voltage.

the fundamental tone, showing that the high-frequency carrier helps to attenuate the IMD components as expected [68].

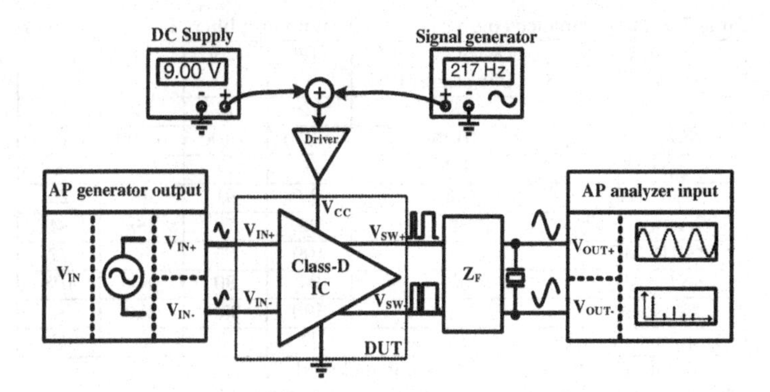

Fig. 7.22 PS-IMD test bench of CDA for PZ speakers.

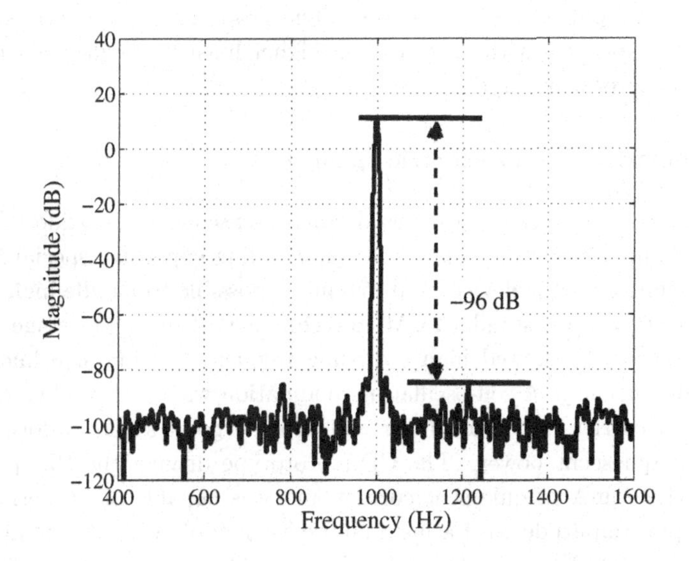

Fig. 7.23 Measured PS-IMD frequency spectrum of CDA for PZ speakers.

To quantify the maximum SPL of a typical PZ speaker driven by the proposed amplifier, an 18 V_{PP} output signal at 2 kHz was used to measure the SPL. The obtained SPL was 96 dB at 10 cm, producing a comparable SPL to EM speakers but with less power consumption.

At the moment of writing, other CDAs for PZ speakers could not be found in the technical literature. Therefore, commercial products were used to compare the performance of the proposed CDA architecture. Table 7.2

Table 7.2 Performance comparison with audio amplifiers for PZ speakers.

Parameter	Stacked-cascode[a] Output stage [128]	[19]	[20]	[21]	[22]
Vout (V_{PP})	18	19	14	20	14
THD+N (%)	**0.025**	0.070	0.100	0.100	0.080
I_Q (mA)	**0.7**	4	17	13	8
P_Q (mW)	**6.3**	22	61	48	29
Efficiency(%)[b]	**96**	92	72	90	84
PSRR (dB)	90	100	65	–	77
SNR (dB)	95	94	80	80	108
F_s (kHz)	750	300	250	250	–
Amplifier class	D	D	D	D	G

[a]High-voltage supply generation not included.
[b]Estimated for 1kHz signal using apparent power.

summarizes the performance of the proposed CDA and compares it with commercial amplifiers for PZ speakers. The proposed architecture is able to drive PZ speakers with 18 V_{PP} with higher linearity, higher efficiency, and lower power consumption.

7.6 Remarks on piezoelectric speakers

The monolithic implementation used stacked-cascode thick-oxide CMOS transistors in the H-bridge output stage, avoiding expensive special high-voltage semiconductor devices and making it possible to handle high voltages in a low voltage standard CMOS technology. The output stage's low input capacitance allowed high switching frequency to improve linearity with high efficiency. A self-oscillating modulation was used to obviate the need for a carrier signal generator and provide good audio performance using low quiescent power. The CDA prototype driving the PZ speaker consumed 0.7 mA of quiescent current and was capable of delivering 18 V_{PP} output amplitude with a maximum efficiency of 96%. The minimum measured THD+N was 0.025% at 5 V_{RMS}. The prototype occupies an active silicon area of 0.4165 mm^2 in standard CMOS 0.18 μm technology. Compared to other CDAs for PZ speakers, the proposed CDA achieved higher linearity, lower power consumption, and higher efficiency at the moment of publication.

Appendix A

Harmonic Distortion in Open-Loop Class-D Amplifiers

This appendix details the derivation of the pulse-width modulated signal and harmonic distortion in open-loop class-D audio amplifiers, for the particular cases of sawtooth, triangle, sinusoidal, and exponential-shaped carrier waveforms, by means of the analysis of duty cycle variation [54]. It also gives the necessary tools to extend the analysis of distortion in class-D audio amplifiers for any periodic carrier waveform and even multilevel modulation schemes.

The analysis by duty-cycle variation is an alternative method to the classical double Fourier integral analysis [54, 55], to calculate the harmonic spectrum in open-loop class-D audio power amplifiers based on naturally sampled pulse-width modulation. This approach examines the switching process of the amplifier during a few arbitrary cycles of the carrier waveform. The reference audio waveform is assumed to be constant within each carrier cycle, i.e. the frequency of the carrier waveform is much higher than the frequency of the audio waveform ($f_c \gg f_o$), which is usually the case.

Firstly, define the existence of two time variables, x(t) and y(t), who represent the time variation of the carrier and the audio waveforms, respectively. These time variables can be expressed as

$$x(t) = \omega_c t + \theta_c \tag{A.1}$$

and

$$y(t) = \omega_o t + \theta_o \tag{A.2}$$

where ω_c is the carrier angular frequency, θ_c is an arbitrary phase offset angle for the carrier waveform, ω_o is the baseband angular frequency, and θ_o is an arbitrary phase offset angle for the baseband waveform. The two angular frequencies (ω_c and ω_o) may not be multiple of each other.

Secondly, recall that any periodic waveform can be represented in terms of its harmonic components. Then, any periodic pulse-width modulated signal can be written as the summation of its Fourier coefficients (a_m and b_m) as

$$v_{PWM}(t) = \frac{a_0}{2} + \sum_{m=1}^{\infty} (a_m \cos mx + b_m \sin mx) \tag{A.3}$$

where

$$a_m = \frac{1}{\pi} \int_{-\pi}^{\pi} v_{PWM}(t) \cos mx \, dx \tag{A.4}$$

and

$$b_m = \frac{1}{\pi} \int_{-\pi}^{\pi} v_{PWM}(t) \sin mx \, dx. \tag{A.5}$$

The pulse-width modulation analysis by duty cycle variation consists on the calculation of the coefficients a_0, a_m, and b_m, in equations (A.3), (A.4), and (A.5), by integrating the duty cycle of the resulting digitally modulated signal within one cycle of the carrier waveform.

A.1 Pulse-width modulation based on sawtooth carrier waveform

The generation of the pulse-width modulated signal based on sawtooth carrier waveform is shown in Fig. A.1. Observe that one cycle the carrier waveform has been normalized to one period equal to 2π (required for Fourier harmonic analysis), and the audio waveform has been expressed as $M\cos(y)$, where M is the modulation index. Notice that $f_c \gg f_o$, and, as mentioned before, the audio waveform can be considered constant within one cycle of the carrier waveform.

For the next step, it is necessary to find the integration limits in the equations (A.4) and (A.5) for the duration of the duty cycle of the pulse-width modulated signal $V_{PWM}(t)$ within one cycle of the carrier waveform, i.e. $-\pi < x < \pi$. In other words, the lower integration limit is calculated when the value of the pulse-width modulated signal $V_{PWM}(t)$ goes high, i.e. when the duty cycle starts; and the higher integration limit is defined when the value of the pulse-width modulated signal $V_{PWM}(t)$ goes low, i.e. when

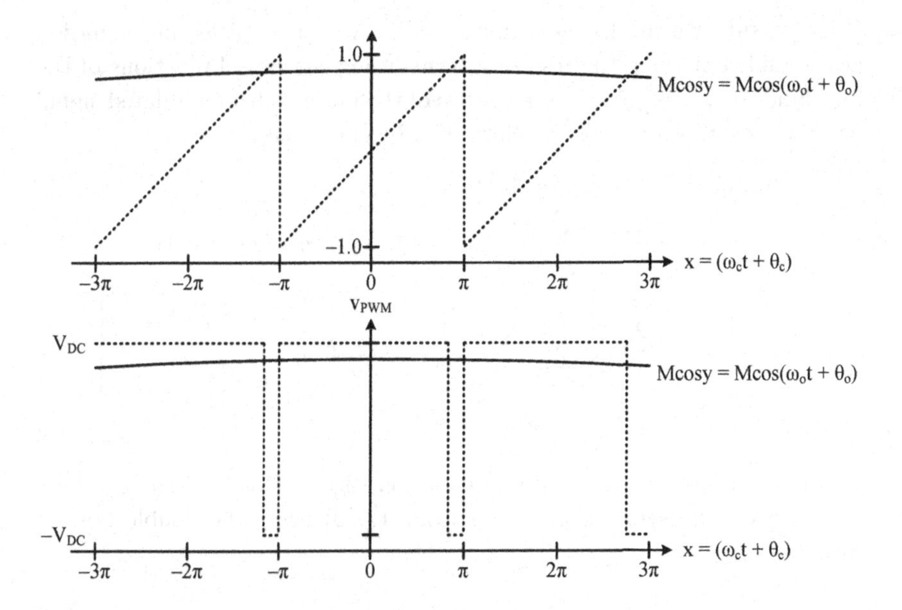

Fig. A.1 Generation of pulse-width modulated signal by comparison of sawtooth carrier wave and audio input wave.

the duty cycle ends. Therefore, the lower integration limit x_L is simply $-\pi$, and the higher integration limit is $x_H = \pi M\cos(y)$ because it is the intersection point of the audio signal $M\cos(y)$ and the carrier waveform, which can be viewed as a line with slope equal to $1/\pi$. Then, equations (A.4) and (A.5) become

$$\begin{aligned} a_m &= \frac{1}{\pi}\int_{-\pi}^{\pi M\cos y} 2V_{DC}\cos mx\,dx \\ &= \frac{2}{m\pi}V_{DC}[\sin(m\pi M\cos y) + \sin m\pi] \end{aligned} \tag{A.6}$$

and

$$\begin{aligned} b_m &= \frac{1}{\pi}\int_{-\pi}^{\pi M\cos y} 2V_{DC}\sin mx\,dx \\ &= \frac{2}{m\pi}V_{DC}[\cos m\pi - \cos(m\pi M\cos y)] \end{aligned} \tag{A.7}$$

when $m \neq 0$. Notice that when $m = 0$,

$$a_0 = 2V_{DC}(1 + M\cos y) \tag{A.8}$$

and

$$b_0 = 0 \tag{A.9}$$

Finally, substituting the equations (A.6), (A.7), and (A.8) into equation (A.3), and combining the resulting terms using the Bessel functions of the first kind $J(\cdot)_{(\cdot)}$ [54], the Fourier series of the pulse-width modulated signal based on sawtooth carrier waveform can be expressed as

$$
\begin{aligned}
V_{PWM}(t) = V_{DC} &+ V_{DC} M \cos y \\
&+ \frac{2}{\pi} V_{DC} \sum_{m=1}^{\infty} \frac{1}{m} [\cos(m\pi) - J_0(m\pi M) \sin mx] \\
&+ \frac{2}{\pi} V_{DC} \sum_{m=1}^{\infty} \sum_{\substack{n=-\infty \\ (n \neq 0)}}^{\infty} \frac{1}{m} J_n(m\pi M) \begin{bmatrix} \sin\left(n\frac{\pi}{2}\right) \cos\gamma \\ -\cos\left(n\frac{\pi}{2}\right) \sin\gamma \end{bmatrix}
\end{aligned}
\tag{A.10}
$$

where γ was defined previously in equation (2.12). Observe that equation (A.10) gives the same result as equation (2.10) using the double Fourier integral analysis.

A.2 Pulse-width modulation based on triangle carrier waveform

The same procedure can be applied to generate the harmonic components of a pulse-width modulated signal based on triangular carrier waveform. The generation of the pulse-width modulated signal $V_{PWM}(t)$ based on triangular carrier waveform in illustrated in Fig. A.2.

Following the same procedure as described above, it is necessary to find the integration limits of the equations (A.4) and (A.5). The intersection point which determines the lower integration limit x_L is giving by the relation

$$M \cos y = -\frac{2}{\pi} x_L - 1 \tag{A.11}$$

because the triangular waveform can be seen as a line with slope -2 / π, shifted down by 1, from -π to 0. On the other hand, the higher integration limit x_H can be calculated by solving

$$M \cos y = \frac{2}{\pi} x_H - 1 \tag{A.12}$$

since in this case, the triangular waveform can be seen as a line with slope 2 / π, also shifted down by 1, from 0 to π.

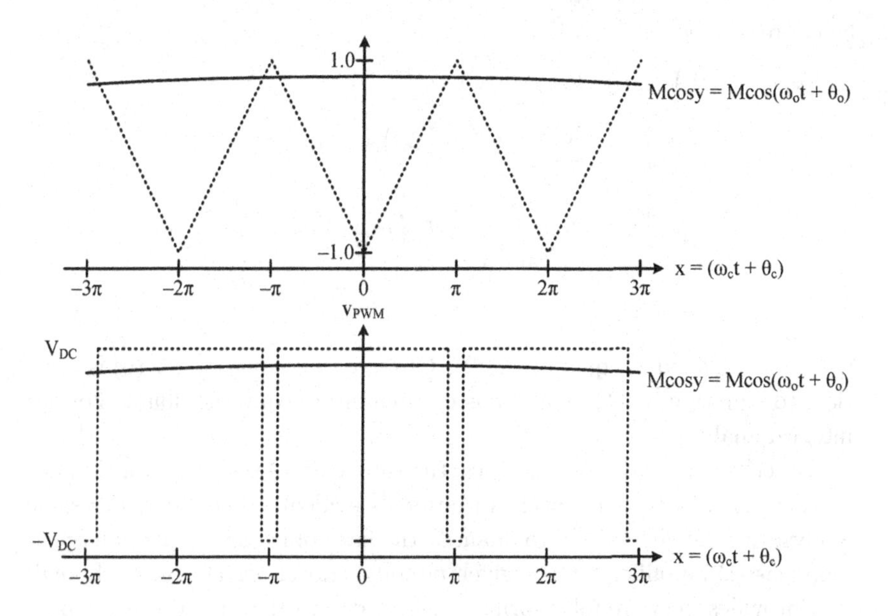

Fig. A.2 Generation of pulse-width modulated signal by comparison of triangle carrier wave and audio input wave.

Therefore, the coefficients in equations (A.4) and (A.5) can be calculated as

$$\begin{aligned} a_m &= \frac{1}{\pi}\int_{-\frac{\pi}{2}(1+M\cos y)}^{\frac{\pi}{2}(1+M\cos y)} 2V_{DC}\cos mx\,dx \\ &= \frac{4}{m\pi}V_{DC}\sin\left(m\frac{\pi}{2}(1+M\cos y)\right) \end{aligned} \tag{A.13}$$

when $m \neq 0$, and

$$b_m = \frac{1}{\pi}\int_{-\frac{\pi}{2}(1+M\cos y)}^{\frac{\pi}{2}(1+M\cos y)} 2V_{DC}\sin mx\,dx = 0 \tag{A.14}$$

because the triangular carrier waveform is an even function. Also, notice that when $m = 0$

$$a_0 = 2V_{DC}(1 + M\cos y). \tag{A.15}$$

Hence, substituting equations (A.13), (A.14), and (A.15) into the general equation (A.3), and after some mathematical manipulation, by employing the Bessel functions of the first kind $J(\cdot)_{(\cdot)}$, the Fourier series of the pulse-width modulated signal based on triangular carrier waveform can be expressed as

$$\begin{aligned} V_{PWM}(t) = V_{DC} &+ V_{DC} M \cos y \\ &+ \frac{4}{\pi} V_{DC} \sum_{m=1}^{\infty} \frac{1}{m} J_0 \left(m \frac{\pi}{2} M \right) \sin \left(m \frac{\pi}{2} \right) \cos mx] \\ &+ \frac{4}{\pi} V_{DC} \sum_{m=1}^{\infty} \sum_{\substack{n=-\infty \\ (n \neq 0)}}^{\infty} \frac{1}{m} J_n \left(m \frac{\pi}{2} M \right) \sin \left([m + n] \frac{\pi}{2} \right) \cos \gamma \end{aligned} \tag{A.16}$$

where γ is defined in equation (2.12). Observe that equation (A.16) is identical to equation (2.11), which was calculated by using the double Fourier integral analysis.

Based on the two previous cases, the simplicity of the pulse-width modulation analysis by duty cycle variation is evident. Therefore, the same analysis can be generalized to quantify the harmonic distortion of an open-loop class-D amplifier for any given periodic carrier waveform. Such analysis provides very useful information in order to determine the required specifications of the carrier waveform generator for a targeted linearity.

A.3 Pulse-width modulation based on band-limited carrier waveforms

The pulse-width modulation based on both, sawtooth and triangle, carrier signals assumes perfect waveforms. In reality, since these carrier waveforms have infinite bandwidth, as shown in equations (2.13) and (2.14), the non-ideal carrier waveforms produces unwanted baseband harmonic distortion at the output of the class-D audio power amplifier. The amount of harmonic distortion can be quantified by analyzing the pulse-width modulated signal by duty cycle variation.

In general, the coefficients a_m and b_m in equation (A.3) can be calculated by evaluating the integrals expressed in equations (A.4) and (A.5) from x_L to x_H as

$$a_m = \frac{1}{\pi} \int_{x_L}^{x_H} v_{PWM}(t) \cos mx \, dx \tag{A.17}$$

and

$$b_m = \frac{1}{\pi} \int_{x_L}^{x_H} v_{PWM}(t) \sin mx \, dx \tag{A.18}$$

within one period of the carrier waveform, i.e. $-\pi < x < \pi$, where the integration limits can be found by solving the equations

$$M \cos y = \frac{1}{2} - \frac{1}{\pi} \sum_{k=1}^{\infty} \frac{1}{k} \sin k x_L, \tag{A.19}$$

$$M \cos y = \frac{1}{2} - \frac{1}{\pi} \sum_{k=1}^{\infty} \frac{1}{k} \sin k x_H \tag{A.20}$$

for a band-limited sawtooth carrier waveform, and

$$M \cos y = \frac{8}{\pi^2} \sum_{k=1,3,5,\ldots}^{\infty} \frac{(-1)^{(k-1)/2}}{k^2} \sin k x_L, \tag{A.21}$$

$$M \cos y = \frac{8}{\pi^2} \sum_{k=1,3,5,\ldots}^{\infty} \frac{(-1)^{(k-1)/2}}{k^2} \sin k x_H \tag{A.22}$$

for a band-limited triangle carrier waveform.

Unfortunately, the evaluation of the integrals expressed in equations (A.17) and (A.18), for $1 < k < \infty$, must be done numerically because they cannot be expressed in a closed-form expression. However, when there is only one harmonic component in the carrier waveform, $k = 1$, the production of the pulse-width modulated signal is based on a pure sinusoidal carrier waveform, and its solution can be expressed in closed form. For example, Fig. A.3 shows the generation of the pulse-width modulated signal $V_{PWM}(t)$ when the triangular carrier waveform contains only one harmonic component.

For this particular case, the integration limits can be found by solving

$$M \cos y = -\frac{8}{\pi^2} \cos x_{L,H} \tag{A.23}$$

for $k = 1$ in equations (A.21) and (A.22), as

$$x_{L,H} = \mp \arccos\left(-\frac{\pi^2}{8} M \cos y\right) \tag{A.24}$$

and the resulting Fourier series is the pulse-width modulated signal

$$\begin{aligned} v_{PWM}(t) &= 2V_{DC} \arccos\left(\arcsin\left[2\sum_{n=1}^{\infty} \sin\left(n\frac{\pi}{2}\right) J_n\left(-\frac{1}{8}\pi^2 M\right) \cos ny\right]\right) \\ &+ \frac{4}{\pi} V_{DC} \sum_{m=1}^{\infty} \sin\left(m \arccos\left[-\frac{1}{8}\pi^2 M \cos y\right]\right) \cos mx \end{aligned} \tag{A.25}$$

which has already been presented in equation (2.15), and is repeated here for completeness.

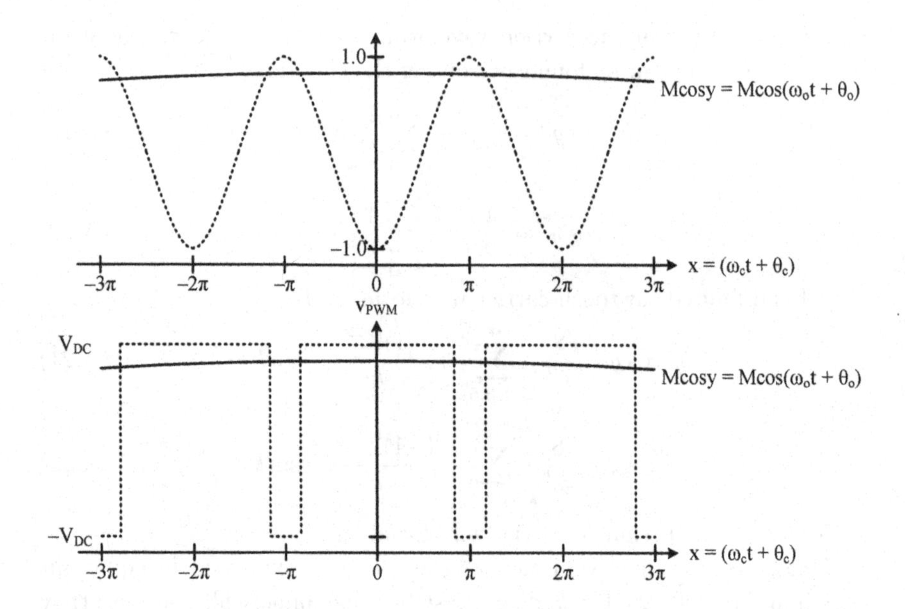

Fig. A.3 Generation of pulse-width modulated signal by comparison of cosine carrier wave and audio input wave.

A.4 Pulse-width modulation based on exponential-shaped carrier waveforms

The analysis of pulse-width modulated signals can be extended to the set of exponential-shaped carrier waveforms defined by equation (2.16). For example, the generation of the pulse-width modulated signal $V_{PWM}(t)$ with a particular exponential-shaped carrier waveform is shown in Fig. A.4.

The calculation of the Fourier coefficients for this particular modulation also requires to find the integration limits in equations (A.17) and (A.18) for the coefficients a_0, a_m, and b_m. Therefore, the intersection points $x_{L,H}$ can be found by equating the sinusoidal audio signal (M cos y) with the exponential function in equation (2.16) for the subintervals $-\pi < x < 0$ and $0 < x < \pi$. The resulting integration limits are

$$x_L = -\pi - t_0 \ln\left(\frac{M}{2V_{DC}}(1 - N_e)\cos y + \frac{(1 + N_e)}{2}\right), \tag{A.26}$$

$$x_H = -t_0 \ln\left(1 - \frac{1}{2}(1 - N_e)\left[\frac{M}{V_{DC}}\cos y + 1\right]\right) \tag{A.27}$$

and the resulting pulse-width modulated signal $V_{PWM}(t)$ is expressed in

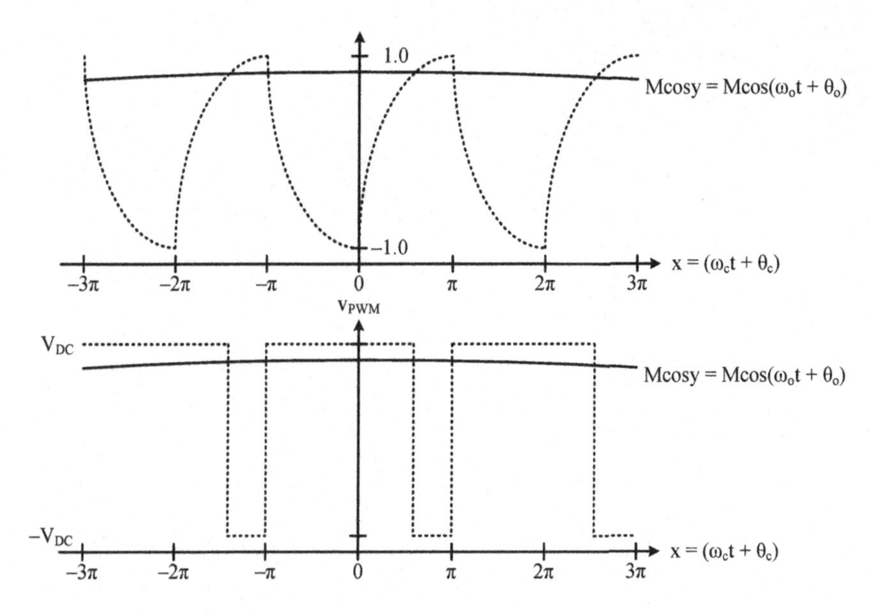

Fig. A.4 Generation of pulse-width modulated signal by comparison of exponential-shaped carrier wave and audio input wave.

equation (2.18) with coefficients a_0, a_m, and b_m specified in equations (2.19), (2.20), and (2.21).

As mentioned before, the analysis of harmonic distortion in class-D audio power amplifiers can be extended to any periodic carrier waveform and even to architectures with multilevel pulse-width modulation.

Appendix B

Fundamentals of Sliding Mode Control

This appendix presents the fundamentals of sliding mode control (SMC) theory. It begins with an introductory example to illustrate its principles of operation, and to highlight its main characteristics. Additionally, a formal description of the sliding mode controller, and the switching function, is given. Furthermore, the analysis of stability, based on the Lyapunov function approach and the equivalent control approach, is explained. Finally, the derivation of the switching function and the stability proof, for the particular cases of the second-order low-pass filter employed in the design of the systems described in this book, are detailed.

B.1 An introductory example to sliding mode control

The first developments of sliding mode control occurred in the 1950s as a consequence of the analysis of discontinuous variable structure systems (VSS). A variable structure system consists of a set of continuous subsystems together with a switching logic. Therefore, the variable structure control (VSC) with sliding modes consists on selecting the parameters of each one of these substructures to define the switching logic of the system. The most outstanding feature of variable structure control is its ability to result in very robust control systems, insensitive to parametric uncertainty, and external disturbances [92, 130]

The basic idea of variable structure control with sliding modes, or simply sliding mode control, can be illustrated by analyzing the second order system shown in Fig. B.1. The system can be expressed in terms of its state variables as

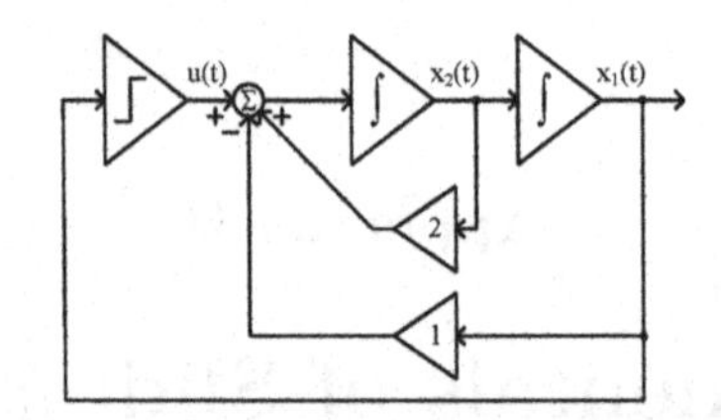

Fig. B.1 Model of a simple variable structure system.

$$\begin{pmatrix} \frac{d}{dt}x_1(t) \\ \frac{d}{dt}x_2(t) \end{pmatrix} = \begin{pmatrix} 0 & 1 \\ -1 & 2 \end{pmatrix} \begin{pmatrix} x_1(t) \\ x_2(t) \end{pmatrix} + \begin{pmatrix} 0 \\ 1 \end{pmatrix} u(t) \tag{B.1}$$

where

$$u(t) = \begin{cases} 4 & \text{when } s(x_1, x_2, t) > 0, \\ -4 & \text{when } s(x_1, x_2, t) < 0 \end{cases} \tag{B.2}$$

and $s(x_1, x_2, t)$, defined as

$$s(x_1, x_2, t) = x_1(t) \left(\frac{1}{2} x_1(t) \ + \ x_2(t) \right) \tag{B.3}$$

represents the switching function, which will be defined later in the appendix. Therefore, the second-order system, in equation (B.1), is analytically defined in two regions of the phase plane, i.e. the $x_1 - x_2$ plane, by two different mathematical models. The first model, when $s(x_1, x_2, t) < 0$, is

$$\begin{pmatrix} \frac{d}{dt}x_1(t) \\ \frac{d}{dt}x_2(t) \end{pmatrix} = \begin{pmatrix} 0 & 1 \\ -5 & 2 \end{pmatrix} \begin{pmatrix} x_1(t) \\ x_2(t) \end{pmatrix} \tag{B.4}$$

and the second model, when $s(x_1, x_2, t) \ > \ 0$, is

$$\begin{pmatrix} \frac{d}{dt}x_1(t) \\ \frac{d}{dt}x_2(t) \end{pmatrix} = \begin{pmatrix} 0 & 1 \\ 3 & 2 \end{pmatrix} \begin{pmatrix} x_1(t) \\ x_2(t) \end{pmatrix}. \tag{B.5}$$

The phase portraits, i.e. the trajectories of the state-space variables in the phase plane for different initial conditions, for the models in equations (B.4) and (B.5) are shown in Fig. B.2 and Fig. B.3, respectively. Figure B.2 corresponds to the state-space model in equation (B.4) and represents the first region of operation, i.e. region I. Observe that the equilibrium point is an unstable focus [130], i.e. positive eigenvalues with imaginary part, at the origin.

On the other hand, the second region of operation, or region II, is represented by the phase portrait, of the state space model expressed in equation (B.5), in Fig. B.3. Notice that, in this case, its equilibrium point, at the origin, is a saddle point [130], i.e. one positive and one negative real eigenvalues, and therefore, it is stable for only one trajectory.

The variable $s(x_1, x_2, t)$ in equation (B.3) describes lines dividing the phase plane into the regions of operation where $s(x_1, x_2, t)$ has different sign. Such lines are called switching lines and $s(x_1, x_2, t)$ is called the switching function. The switching lines occur whenever $s(x_1, x_2, t) = 0$ and are known as the switching surfaces. Hence, the feedback control u(t) switches according to the sign of $s(x_1, x_2, t)$. For example, the switching function in equation (B.3) defines the phase portrait, of the second-order system in equation (B.1), as illustrated in Fig. B.4. The phase plane is divided into regions of operation, each one of them linked to the state-space systems in equations (B.4) and (B.5). The switching function controls the switching logic to stabilize the system for any given initial condition.

The phase trajectories, plotted in the phase portrait of Fig. B.4, correspond to the two modes of operation of the system. The first part is the reaching mode, also called nonsliding mode, in which a trajectory starting at any initial condition moves toward a switching line and reaches the line in finite time. The second part is the sliding mode, in which the trajectory asymptotically tends to the origin of the phase plane. This displacement is called sliding because in the ideal case, the system switches at infinite frequency, causing a sliding behavior of the particular trajectory.

During the control process, the variable structure system, in equation (B.1), varies from one structure to another, thus earning the name variable structure control. The control is also called sliding mode control to emphasize the important role of sliding mode [92, 130].

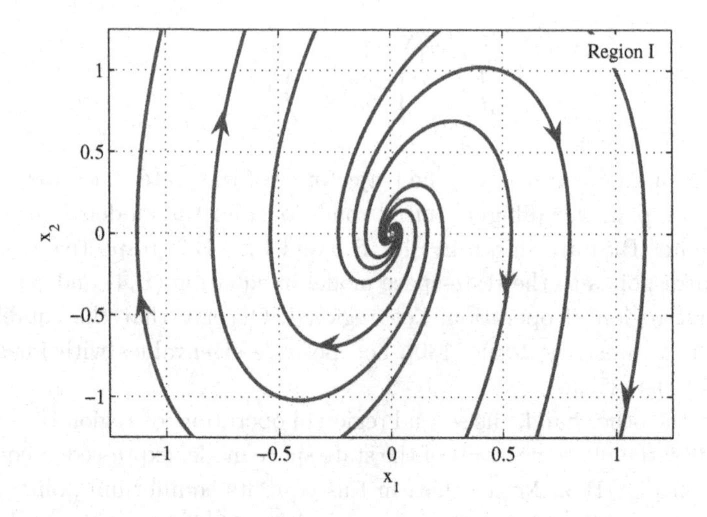

Fig. B.2 Phase portrait of the second-order system in equation (B.1) for Region I when $s(x_1, x_2, t) < 0$.

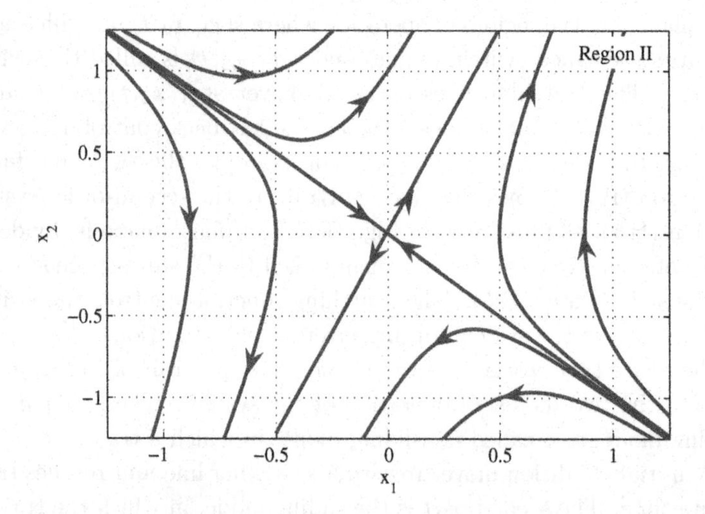

Fig. B.3 Phase portrait of the second-order system in equation (B.1) for Region II when $s(x_1, x_2, t) > 0$.

B.2 Sliding mode controller

The switching function represents the sliding mode controller, i.e. the control law, of a variable structure system. Hence, if the variable structure

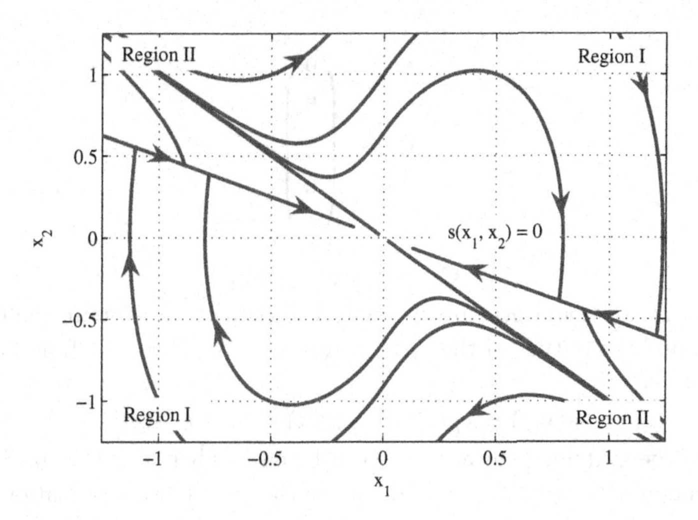

Fig. B.4 Phase portrait of the second-order system in equation (B.1) with sliding mode.

system is expressed in the controllable canonical form as

$$\frac{d}{dt}\mathbf{x}(t) = \mathbf{A}\mathbf{x}(t) + \mathbf{B}u(t), \tag{B.6}$$

$$y(t) = \mathbf{C}\mathbf{x}(t) \tag{B.7}$$

where

$$\mathbf{x}(t) = \begin{pmatrix} x_1(t) \\ x_2(t) \\ \vdots \\ x_{n-1}(t) \\ x_n(t) \end{pmatrix}, \tag{B.8}$$

$$\mathbf{A} = \begin{pmatrix} 0 & 1 & 0 & \cdots & 0 \\ 0 & 0 & 1 & \cdots & 0 \\ \vdots & \vdots & \vdots & \ddots & \vdots \\ 0 & 0 & 0 & \cdots & 1 \\ -a_1 & -a_2 & -a_3 & \cdots & -a_n \end{pmatrix}, \tag{B.9}$$

$$\mathbf{B} = \begin{pmatrix} 0 \\ 0 \\ \vdots \\ 0 \\ 1 \end{pmatrix}, \tag{B.10}$$

$$\mathbf{C} = \begin{pmatrix} c_1 \; c_2 \cdots c_n \end{pmatrix} \tag{B.11}$$

and $x_n(t)$, $u(t)$, and $y(t)$ are the state variables of the system, the control input, and the output of the system, respectively. Then, the function

$$\mathbf{s}(\mathbf{x}, t) = k_1 x_1(t) + k_2 x_2(t) + \cdots + k_n x_n(t) \tag{B.12}$$

defines the switching surfaces in the nth space, when $\mathbf{s}(\mathbf{x}, t) = 0$. The coefficients in the switching function define the characteristic equation of the sliding mode if the system model is described in the controllable canonical form [92,130].

In the same way, the control law can be designed such that the output of the system $y(t)$ asymptotically tracks a reference signal $r(t)$. Therefore, if the variable structure system is rewritten with

$$\begin{pmatrix} \dot{e}_1(t) \\ \dot{e}_2(t) \\ \vdots \\ \dot{e}_{n-1}(t) \end{pmatrix} = \begin{pmatrix} 0 & 1 & 0 & \cdots & 0 \\ 0 & 0 & 1 & \cdots & 0 \\ \vdots & \vdots & \vdots & \ddots & \vdots \\ 0 & 0 & 0 & \cdots & 1 \end{pmatrix} \begin{pmatrix} e_1(t) \\ e_2(t) \\ \vdots \\ e_n(t) \end{pmatrix} \tag{B.13}$$

where $e_1(t) = r(t) - y(t)$ is the error function, $e_n(t)$ is the control input, and n is the order of the system to be controlled. The control input, defined in equation (B.14), is the linear combination of all canonical state variables [130], and whose coefficients are chosen in such way that the polynomial, in equation (B.15), meets the Hurwitz criterion, i.e. all its roots have negative real part.

$$e_n(t) = -[k_1 e_1(t) + k_2 e_2(t) + \cdots + k_{n-1} e_{n-1}(t)], \tag{B.14}$$

$$P(s) = k_n s^{n-1} + k_{n-1} s^{n-2} + \cdots + k_1 \tag{B.15}$$

Then, the switching function in equation (B.16) represents the $(n-1)$-dimensional surface where the points of discontinuity merge.

$$\mathbf{s}(\mathbf{e}, t) = k_1 e_1(t) + k_2 e_2(t) + \cdots + k_{n-1} e_{n-1}(t) + k_n e_n(t) = 0. \tag{B.16}$$

B.3 Stability analysis

Variable structure systems operating under sliding mode control consist of two parts, the reaching mode and the sliding mode. Therefore, the analysis of stability must demonstrate that (1) the trajectory of a given state moves toward and reaches the sliding surface, and (2) the state asymptotically tends to the equilibrium point of the system.

B.3.1 *Reaching mode condition*

The reaching mode condition can be analyzed by employing the Lyapunov function approach. Hence, by choosing the Lyapunov function candidate

$$\mathbf{v}(\mathbf{x}, t) = \frac{1}{2}\mathbf{s}^T(\mathbf{x}, t)\mathbf{s}(\mathbf{x}, t) \tag{B.17}$$

a global reaching condition is given by

$$\frac{d}{dt}\mathbf{v}(\mathbf{x}, t) < 0 \tag{B.18}$$

when $\mathbf{s}(\mathbf{x},\ t) \neq 0$ [130].

B.3.2 *Sliding mode condition*

The convergence of a variable structure system to its equilibrium point, also called sliding equilibrium point or quasi-equilibrium point [94], can be found by analyzing the qualitative behavior [130], i.e. calculating the eigenvalues, of the equivalent variable structure system when

$$\frac{d}{dt}\mathbf{x}(t) = \mathbf{A}\mathbf{x}(t) + \mathbf{B}u_{eq}(t) = 0, \tag{B.19}$$

$$\mathbf{s}(\mathbf{x}, t) = 0 \tag{B.20}$$

where $u_{eq}(t)$ is the equivalent control input that describes the dynamics of the sliding mode as the average value of the discontinuous input $u(t)$ [94]. Hence, if the switching function $\mathbf{s}(\mathbf{x},\ t)$ is expressed in terms of the state variables as

$$\mathbf{s}(\mathbf{x}, t) = \mathbf{D}(\mathbf{x}, t) + \mathbf{E}(\mathbf{x}, t)u(t) \tag{B.21}$$

then, the equivalent control can be found when the state trajectory stays on the switching surface $\mathbf{s}(\mathbf{x}, \mathrm{t}) = 0$. Therefore, differentiating $\mathbf{s}(\mathbf{x}, t)$ with respect to time gives

$$\frac{d}{dt}\mathbf{s}(\mathbf{x},t) = \frac{\partial}{\partial \mathbf{x}}\frac{d}{dt}\mathbf{D}(\mathbf{x},t) + \frac{\partial}{\partial \mathbf{x}}\frac{d}{dt}\mathbf{E}(\mathbf{x},t)u(t) \tag{B.22}$$

and solving equation (B.22) for $u(t)$ yields the equivalent control input $u_{eq}(\mathbf{x}, t)$ as

$$\mathbf{u_{eq}}(\mathbf{x},t) = -\left(\frac{\partial}{\partial \mathbf{x}}\frac{d}{dt}\mathbf{E}(\mathbf{x},t)\right)^{-1}\frac{\partial}{\partial \mathbf{x}}\frac{d}{dt}\mathbf{D}(\mathbf{x},t). \tag{B.23}$$

B.4 Practical derivation of the switching function and stability analysis

If the variable structure system, as described in previous chapters, is defined by the second-order state-space system given by

$$\begin{pmatrix} \frac{d}{dt}i_L(t) \\ \frac{d}{dt}v_C(t) \end{pmatrix} = \begin{pmatrix} 0 & -\frac{1}{L} \\ \frac{1}{C} & -\frac{1}{CR} \end{pmatrix} \begin{pmatrix} i_L(t) \\ v_C(t) \end{pmatrix} + \begin{pmatrix} \frac{1}{L} \\ 0 \end{pmatrix} u(t) \tag{B.24}$$

with an error function $e_1(t) = v_{REF}(t) - v_C(t)$, then, from equations (B.13) and (B.14), we have

$$\frac{d}{dt}e_1(t) = e_2(t), \tag{B.25}$$

$$e_2(t) = -k_1 e_1(t) \tag{B.26}$$

and the switching function $s(e_1, e_2, t)$, from equation (B.16), is defined as

$$s(e_1, e_2, t) = k_1 e_1(t) + k_2 e_2(t) \tag{B.27}$$

where k_1 and k_2 must be chosen such that the polynomial $P(s) = k_2 s + k_1$, from equation (B.15), is Hurwitz. Therefore, the control input u(t) switches according to

$$u(t) = \begin{cases} v_{DD} & \text{when } s(e_1, e_2, t) > 0, \\ v_{SS} & \text{when } s(e_1, e_2, t) < 0. \end{cases} \tag{B.28}$$

Hence, the switching function in equation (B.27) can be rewritten as a function of the state-space variables as

$$\begin{aligned} s(e_1, e_2, t) &= e_1(t) + \alpha e_2(t) \\ &= v_{REF}(t) - v_C(t) - \alpha\frac{d}{dt}v_C(t) \end{aligned} \tag{B.29}$$

and the derivative of the switching function, from equation (B.21), is

$$\dot{s}(e_1, e_2, t) = \frac{1}{C}\left(\frac{\alpha}{RC} - 1\right) i_L(t) - \left[\frac{1}{RC}\left(\frac{\alpha}{RC} - 1\right) - \frac{\alpha}{LC}\right] v_C(t) - \frac{\alpha}{LC} u(t). \quad \text{(B.30)}$$

The analysis of stability based on the Lyapunov function approach assumes the control signal u(t) can be decomposed into two parts

$$u(t) = u_{eq}(t) + u_{nl}(t) \quad \text{(B.31)}$$

where $u_{eq}(t)$ is the equivalent control input, and $u_{nl}(t)$ is the nonlinear switching function, i.e. the high-frequency component. Therefore, the equivalent control input, defined in equation (B.23), for this particular case is

$$u_{eq}(t) = \left(\frac{L}{\alpha}\left[\frac{\alpha}{RC} - 1\right] 1 - \frac{L}{\alpha R}\left[\frac{\alpha}{RC} - 1\right]\right)\begin{pmatrix} i_L(t) \\ v_C(t) \end{pmatrix} \quad \text{(B.32)}$$

hence, substituting equations (B.28) and (B.32) into equation (B.30) yields

$$\dot{s}(e_1, e_2, t) = -\frac{\alpha}{CL} u_{nl}(t). \quad \text{(B.33)}$$

Therefore, the Lyapunov function candidate, from equation (B.17), becomes

$$v(e_1, e_2, t) = \frac{1}{2} s^2(e_1, e_2, t) \quad \text{(B.34)}$$

and the global reaching condition is

$$\begin{aligned} \frac{d}{dt} v(e_1, e_2, t) &= s(e_1, e_2, t)\dot{s}(e_1, e_2, t) \\ &= s(e_1, e_2, t)\left(-\frac{\alpha}{CL} u_{nl}(t)\right) < 0 \end{aligned} \quad \text{(B.35)}$$

when $s(e_1, e_2, t) \neq 0$. Simplifying and rearranging we get

$$s(e_1, e_2, t) u_{nl}(t) > 0. \quad \text{(B.36)}$$

Hence, based on equations (B.28) and (B.31), when $s(e_1, e_2, t) > 0$, then $u(t) = v_{DD}$ and thus $v_{DD} = u_{eq} + u_{nl}$, therefore, if $v_{DD} - u_{eq} > 0$, it implies that $u_{nl} > 0$ and

$$[s(e_1, e_2, t)][u_{nl}(t)] > 0. \quad \text{(B.37)}$$

for $s(e_1, e_2, t) > 0$. On the other hand, when $s(e_1, e_2, t) < 0$, then $u(t) = v_{SS}$, so $v_{SS} = u_{eq} + u_{nl}$, this implies that if $v_{SS} - u_{eq} < 0$, therefore $u_{nl} < 0$ and

$$[-s(e_1, e_2, t)][-u_{nl}(t)] > 0 \quad \text{(B.38)}$$

for $s(e_1, e_2, t) < 0$. Then, if $v_{SS} < u_{eq} < v_{DD}$ holds, the control law ensures the reaching condition. Since we know that u_{eq} is the low-frequency average signal that tracks the reference input v_{ref}, then the last inequality is true.

On the other hand, the sliding mode condition can be proven if the sliding equilibrium point of the equivalent control system is found, and its eigenvalues have negative real part. Therefore, the equivalent input control input in equation (B.32) is substituted in the state-space model in equation (B.24) as

$$\begin{pmatrix} \frac{d}{dt} i_L(t) \\ \frac{d}{dt} v_C(t) \end{pmatrix} = \begin{pmatrix} \frac{1}{\alpha}\left(\frac{\alpha}{RC} - 1\right) & -\frac{1}{\alpha R}\left(\frac{\alpha}{RC} - 1\right) \\ \frac{1}{C} & -\frac{1}{CR} \end{pmatrix} \begin{pmatrix} i_L(t) \\ v_C(t) \end{pmatrix}. \quad \text{(B.39)}$$

Then, as shown in equations (B.19) and (B.20), if the resulting equivalent control system, along with the switching function are solved, when they are equal to zero, the sliding equilibrium point yields

$$[v_C(t), i_L(t)] = \left[v_{REF}(t), \frac{v_{REF}(t)}{R}\right]. \quad \text{(B.40)}$$

The sliding equilibrium point corresponds to the desired voltage $v_{REF}(t)$ at the output second-order low-pass filter. Assuming that $v_C(t) = v_{OUT}(t)$, the sliding mode controller will track the trajectory of the input signal $v_{REF}(t)$. Similarly, the value of the inductor current $i_L(t)$ will be defined by the output voltage divided by the resistive load.

The value of the eigenvalues in the equivalent control model can be calculated to show that the system converges to the sliding equilibrium point. Therefore, solving for $v_C(t)$ in equation (B.29), when $s(e_1, e_2, t) = 0$, and substituting into the equivalent control model expressed in equation (B.39), the eigenvalues (λ) of the equivalent system are

$$\lambda_{1,2} = \left(-\frac{1}{\alpha}, -\frac{1}{RC}\right). \quad \text{(B.41)}$$

Thus, the system is asymptotically stable since its sliding equilibrium point is a node whose eigenvalues are real and negative, for $\alpha > 0$.

Furthermore, the final value theorem (FVT) can be used in order to calculate the steady-state of the model to verify that system under sliding mode is in fact a tracking system. In general, the final value of a given system y(t) can be determined as

$$\lim_{t \to \infty} y(t) = \lim_{s \to 0} sY(s). \quad \text{(B.42)}$$

The transfer function of the equivalent control model, resulting from the combination of equations (B.29) and (B.39), is

$$\frac{V_{OUT}(s)}{V_{REF}(s)} = \frac{1}{(\alpha s + 1)(RCs + 1)} \tag{B.43}$$

which agrees with the results given in equation (B.41) for the eigenvalues of the equivalent control model.

Applying the final value theorem to equation (B.43) with a step input of value v_{STEP} to the system we have

$$\begin{aligned}\lim_{t\to\infty} v_{OUT}(t) &= \lim_{s\to 0} sV_{OUT}(s) \\ &= \lim_{s\to 0}\left(\frac{s}{(\alpha s + 1)(RCs + 1)}\right)\left(\frac{V_{STEP}}{s}\right) \\ &= v_{STEP}.\end{aligned} \tag{B.44}$$

Hence, the equivalent control model tracks the input step signal v_{STEP}.

B.5 Equivalent control model and stability analysis for integral sliding mode

This section includes the derivation of the equivalent control model and the stability analysis for the class-D amplifier operating under integral sliding mode control. First, the state-space model corresponding to the 2nd-order LPF of the class-D amplifier can be expressed as shown in B.24, where $v_C(t)$ is the voltage across the capacitor C, $i_L(t)$ is the current across the inductor L, R represents the speaker resistance, and $u(t)$ is the binary modulated signal generated by the ISMC. This control signal makes the output stage of the audio amplifier to switch between the supply voltage and ground according to the sign of the switching function in Eq. (6.18) as,

$$u(t) = \begin{cases} V_{DD} & \text{when } s(v_e, v_i, t) > 0, \\ 0 & \text{when } s(v_e, v_i, t) < 0. \end{cases} \tag{B.45}$$

The equivalent control approach [92] decomposes the discontinuous control function $u(t)$ as the sum of a high-frequency term, $u_o(t)$, and a low-frequency component, $u_{eq}(t)$, where the latter can be viewed as the average value of the discontinuous function, i.e. the equivalent control input.

$$u(t) = u_{eq}(t) + u_o(t). \tag{B.46}$$

Therefore, it is required to calculate the input $u_{eq}(t)$ such that the states trajectories stay on the switching surface, i.e. $s(v_e, v_i, t) = 0$. A necessary condition is that $\dot{s}(v_e, v_i, t) = 0$. Then, differentiating Eq. (6.18) with respect to time and solving for $u(t)$ we can obtain the equivalent control inputs as

$$u_{eq}(t) = (1 - L\beta)v_C(t) \; + \; L\beta v_{in}(t). \tag{B.47}$$

Substituting Eq. (B.47) into the state-space model in Eq. (B.24) we can obtain the general equivalent state-space model given by,

$$\begin{pmatrix} \frac{d}{dt}i_L(t) \\ \frac{d}{dt}v_C(t) \end{pmatrix} = \begin{pmatrix} 0 & -\beta \\ \frac{1}{C} & -\frac{1}{CR} \end{pmatrix} \begin{pmatrix} i_L(t) \\ v_C(t) \end{pmatrix} + \begin{pmatrix} \beta \\ 0 \end{pmatrix} v_{in}(t). \tag{B.48}$$

By definition [92], the sliding equilibrium point of the equivalent state-space model in Eq. (B.48) can be obtained if

$$\begin{pmatrix} \frac{d}{dt}i_L \\ \frac{d}{dt}v_C \end{pmatrix} = 0 \tag{B.49}$$

when $s(v_e, v_i, t) = 0$. Hence, the sliding equilibrium of the proposed class-D audio amplifier is given by

$$v_C(t) = v_{in}, \tag{B.50}$$

$$i_L = \frac{v_{in}}{R}. \tag{B.51}$$

The sliding equilibrium point tracks the value of the input voltage, i.e. $v_{out}(t)$ follows $v_{in}(t)$. Similarly, the value of the output currents is defined by the ratio of the output voltage and the speaker resistance. Furthermore, the eigenvalues of the equivalent state-space model in Eq. (B.48) correspond to an stable focus since their values are complex with real negative part.

Finally, the final value theorem (FVT) [92] can be used in order to calculate the steady-state response of the equivalent control state-space model to verify that the class-D audio amplifier operating under sliding mode is in fact a tracking system.

The transfer function of the equivalent control model, is

$$\frac{V_{OUT}(s)}{V_{IN}(s)} = \frac{\beta/C}{s^2 + s/CR + \beta/C}. \tag{B.52}$$

Applying the final value theorem to Eq. (B.52) with a step input of value $V_{IN}(s)$ to the system we have

$$\begin{aligned} \lim_{t\to\infty} v_{out}(t) &= \lim_{s\to 0} sV_{OUT}(s) \\ &= v_{in}(t). \end{aligned} \tag{B.53}$$

Hence, the equivalent control model tracks the input step signal $v_{in}(t)$.

Appendix C

Switching Frequency of Class-D Amplifiers with Sliding Mode Control

This appendix derives the expressions for the calculation of the switching frequency in class-D audio power amplifiers operating under sliding mode. The analysis is done by assuming a steady state operation of the amplifier and a constant load R. The derivations are calculated for two different cases: (1) when an amplifier is operating under ideal sliding mode, and (2) when the amplifier is based on a lossy sliding mode. Also, for this analysis, the output stage and second-order filter are assuming to be described by the second-order state space system defined in Appendix B as equation (B.24).

C.1 Class-D amplifier operating under ideal sliding mode

A magnified view of the class-D audio power amplifier operating under ideal sliding mode has been shown in Fig. 6.18. The time duration of subintervals Δt_1 and Δt_2 can be calculated as

$$\Delta t_1 = \frac{2\kappa}{\frac{d}{dt}s(e_1,t)} = \frac{2\kappa}{\frac{d}{dt}\left(e_1(t) + \alpha\frac{d}{dt}e_1(t)\right)} \tag{C.1}$$

when $V_{IN} = V_{PWM-} = 0$, and

$$\Delta t_2 = \frac{-2\kappa}{\frac{d}{dt}s(e_1,t)} = \frac{-2\kappa}{\frac{d}{dt}\left(e_1(t) + \alpha\frac{d}{dt}e_1(t)\right)} \tag{C.2}$$

when $V_{IN} = V_{PWM+} = V_{DD}$. Therefore, the time period for one cycle of operation is giving by

$$T_{s,ideal} = \Delta t_1 + \Delta t_2 = \frac{2\kappa}{\varphi} - \frac{2\kappa}{\varphi - \frac{\alpha}{LC}v_{DD}} \tag{C.3}$$

where

$$\varphi = \frac{1}{C}\left(\frac{\alpha}{RC} - 1\right) i_L - \left(\frac{1}{RC}\left(\frac{\alpha}{RC} - 1\right) - \frac{\alpha}{LC}\right) v_C \tag{C.4}$$

and the switching frequency is simply the inverse of equation (C.3)

$$f_{s,ideal} = \frac{\varphi\left(\varphi - \frac{\alpha}{LC} v_{DD}\right)}{-2\kappa \frac{\alpha}{LC} v_{DD}}. \tag{C.5}$$

Hence, if we substitute the value of the derivative constant by $\alpha = RC \approx 5.625 \cdot 10^{-6}$, the equation (C.5) reduces to

$$f_{s,ideal} = \frac{1}{2\kappa} \frac{R}{L} v_C \left(1 - \frac{v_C}{v_{DD}}\right) \tag{C.6}$$

as expressed previously in equation (6.12).

C.2 Class-D amplifier operating under lossy sliding mode

In practice, the sliding mode controller is implemented with a lossy differentiator to bound the bandwidth of the class-D audio power amplifier and to limit the high-frequency noise. Hence, the switching frequency of the class-D audio power amplifier operating under a lossy switching function becomes inversely proportional to the frequency of the extra pole added.

The derivation of the expression of the switching frequency with a lossy-differentiator follows the same procedure as the ideal case, but, in this case, considering the lossy-switching function $s(e_1, t)$. Firstly, the lossy-switching function, in equation (6.4), can be rewritten, using the partial-fraction expansion method, as

$$\begin{aligned} S(E_1, s) &= \left[1 + \left(\frac{\alpha s}{\frac{1}{\omega_p} s + 1}\right)\right] E_1(s) = \left[1 + \left(\frac{\alpha s}{\frac{\alpha}{\gamma} s + 1}\right)\right] E_1(s) \\ &= \left[(1 + \gamma) - \frac{\gamma^2}{\alpha}\left(\frac{\alpha}{s + \frac{\gamma}{\alpha}}\right)\right] E_1(s). \end{aligned} \tag{C.7}$$

Then, the lossy-switching function in equation (C.7) can be expressed in the time domain, applying the inverse Laplace transform, as

$$\begin{aligned} s(e_1, t) &= L^{-1}[S(E_1, s)] \\ &= (1 + \gamma)e_1(t) - \frac{\gamma^2}{\alpha}\exp\left(-\frac{\gamma}{\alpha}t\right) * e_1(t) \\ &= (1 + \gamma)e_1(t) - \frac{\gamma^2}{\alpha}\int_{-\infty}^{\infty}\exp\left(-\frac{\gamma}{\alpha}(t - \tau)\right)e_1(\tau)\,d\tau \\ &= (1 + \gamma)e_1(t) - \frac{\gamma^2}{\alpha}\exp\left(-\frac{\gamma}{\alpha}t\right)\int_0^t \exp\left(\frac{\gamma}{\alpha}\tau\right)e_1(\tau)\,d\tau \\ &= (1 + \gamma)e_1(t) - \frac{\gamma^2}{\alpha}\exp\left(-\frac{\gamma}{\alpha}t\right)g(t) \end{aligned} \tag{C.8}$$

where

$$\begin{aligned} g(t) = \exp\left(\frac{\gamma}{\alpha}t\right)&\sum_{n = 1}^{\infty}(-1)^{(n - 1)}\left(\frac{\alpha}{\gamma}\right)^n e_1^{(n - 1)}(t) \\ &- \sum_{n = 1}^{\infty}(-1)^{(n - 1)}\left(\frac{\alpha}{\gamma}\right)^n e_1^{(n - 1)}(0) \end{aligned} \tag{C.9}$$

Rearranging terms and simplifying, we have

$$\begin{aligned} s(e_1, t) = (1 + \gamma)e_1(t) & \\ -\sum_{n=1}^{\infty}&(-1)^{(n-1)}\left(\frac{\alpha^{(n-1)}}{\gamma^{(n-2)}}\right)e_1^{(n-1)}(t) \\ +\exp\left(-\frac{\gamma}{\alpha}t\right)\sum_{n=1}^{\infty}&(-1)^{(n-1)}\left(\frac{\alpha^{(n-1)}}{\gamma^{(n-2)}}\right)e_1^{(n-1)}(0). \end{aligned} \tag{C.10}$$

The resulting equation (C.10) is an infinite sum of derivative functions, but can be rewritten, for simplicity, by only taking the first three coefficients in the summation terms, as follows

$$\begin{aligned} s(e_1, t) \approx &\left[e_1(t) + \alpha\frac{d}{dt}e_1(t) - \frac{\alpha^2}{\gamma}\frac{d^2}{dt^2}e_1(t)\right] \\ &+\exp\left(-\frac{\gamma}{\alpha}t\right)\left[\gamma e_1(0) - \alpha\frac{d}{dt}e_1(0) + \frac{\alpha^2}{\gamma}\frac{d^2}{dt^2}e_1(0)\right]. \end{aligned} \tag{C.11}$$

and whose derivative, assuming $\alpha = RC \approx 5.625 \cdot 10^{-6}$, is

$$\begin{aligned} \frac{d}{dt}s(e_1, t) \approx &\frac{R}{L}(v_C - v_{IN}) + \frac{1}{\gamma}\left[\frac{1}{2C}i_L - \frac{R}{L}v_{IN}\right] \\ &- \exp\left(\frac{\gamma}{\alpha}t\right)\left[\frac{\gamma^2}{\alpha}e_1(0) - \gamma\frac{d}{dt}e_1(0) + \alpha\frac{d^2}{dt^2}e_1(0)\right]. \end{aligned} \tag{C.12}$$

However, equation (C.12) depends on the initial condition of the error function and its derivatives, which are unknown. Therefore, the derivative of the lossy-switching function is approximated to

$$\frac{d}{dt}s(e_1,t) \approx \frac{R}{L}(v_C - v_{IN}) + \frac{1}{\gamma}\left[\frac{1}{2C}i_L - \frac{R}{L}v_{IN}\right]. \qquad \text{(C.13)}$$

Hence, the original two subintervals of operation of the class-D audio power amplifier operating under ideal-sliding control are expanded into a total of six subintervals of operation under the lossy-sliding control, as shown in Fig. 6.19. Two of the subintervals, Δt_2 and Δt_5, are related to the derivative of lossy-switching function in equation (C.13), and four of the subintervals, Δt_1, Δt_3, Δt_4, and Δt_6, account for the truncated exponential terms of the lossy-switching function in equation (C.12).

Therefore, subintervals Δt_2 and Δt_5 and are defined as

$$\Delta t_2 = \frac{2\kappa}{\frac{d}{dt}s(e_1,t)} \qquad \text{(C.14)}$$

when $V_{IN} = V_{PWM-} = 0$, and

$$\Delta t_5 = \frac{-2\kappa}{\frac{d}{dt}s(e_1,t)} \qquad \text{(C.15)}$$

when $V_{IN} = V_{PWM+} = V_{DD}$, and subintervals Δt_1, Δt_3, Δt_4, and Δt_6, are approximated by calculating the time that takes to the exponential wave in equation (C.16) to decay down to 1% of its initial value, at t = 0, for the maximum switching frequency in the ideal sliding mode as

$$\exp\left(-\frac{\gamma}{\alpha}t\right) = \exp\left(-\frac{\gamma}{4\alpha f_{s,ideal}k_t}\right) = 0.01 \qquad \text{(C.16)}$$

then, subintervals Δt_1, Δt_3, Δt_4, and Δt_6 can be expressed as

$$\Delta t_1 + \Delta t_3 + \Delta t_4 + \Delta t_6 = 4\Delta t_0 = -4\frac{\alpha}{\gamma}\ln\left(\frac{v_H - \kappa}{\gamma e_1}\right) \qquad \text{(C.17)}$$

where

$$v_H = e_1\gamma\exp(-k_t) + \kappa[1 - \exp(-k_t)] \qquad \text{(C.18)}$$

and k_t, from equation (C.16) is

$$k_t = -\frac{\gamma}{4\alpha\ln(0.01)}\left(\frac{1}{f_{s,ideal}}\right) \qquad \text{(C.19)}$$

as expressed before in equations (6.14), and (6.15).

Finally, the time period for one cycle of operation is giving by

$$\begin{aligned}
T_{s,real} &\approx \Delta t_1 + \Delta t_2 + \Delta t_3 + \Delta t_4 + \Delta t_5 + \Delta t_6 \\
&\approx \Delta t_2 + \Delta t_5 + 4\Delta t_0 \\
&\approx \frac{2\kappa V_{DD}\frac{R}{L}\left(1 + \frac{1}{\gamma}\right)}{\left(\frac{R}{L}v_C + \frac{1}{2\gamma C}i_L\right)\left(V_{DD}\frac{R}{L}\left(1 + \frac{1}{\gamma}\right) - \left(\frac{R}{L}v_C + \frac{1}{2\gamma C}i_L\right)\right)} \\
&\quad - 4\frac{\alpha}{\gamma}\ln\left(\frac{v_H - \kappa}{\gamma e_1}\right) \qquad \text{(C.20)}
\end{aligned}$$

as defined previously in equation (6.13).

Bibliography

[1] D. Self, *Audio Power Amplifier Design Handbook*. Newnes-Elsevier, Oxford, UK (2006).

[2] B. Forejt, *Tutorial T1: Fundamentals of Class-D Amplifier Operation & Design* (2008), Presented at the *IEEE International Solid-State Circuits Conference*, San Francisco, CA, February 2008.

[3] W. M. Leach, *Introduction to Electroacoustics and Audio Amplifier Design*. Kendall/Hunt Publishing Company, Iowa (2003).

[4] E. M. Villchur, Amplifiers, *Audio* **39**, p. 65 (1955).

[5] L. De Forest, The Audion — I: A New Receiver for Wireless Telegraphy, *Scientific American Supplement* **1**, 1665, pp. 348–350 (1907).

[6] L. De Forest, The Audion — II: A New Receiver for Wireless Telegraphy, *Scientific American Supplement* **1**, 1666, pp. 354–356 (1907).

[7] P. J. Baxandall, Transistor Sine-Wave LC Oscillators. Some General Considerations and New Developments, *Proceedings of the IEE — Part B: Electronic and Communication Engineering* **106**, 16, pp. 748–758 (1959).

[8] S. E. Keith, D. S. Michaud and V. Chiu, Evaluating the Maximum Playback Sound Levels from Portable Digital Audio Players, *Journal of the Acoustical Society of America* **123**, 6, pp. 4227–4237 (2008).

[9] S. Oksanen, M. Hiipakka and V. Sivonen, Estimating Individual Sound Pressure Levels at the Eardrum in Music Playback over Insert Headphones, in *Audio Engineering Society Conference: Music Induced Hearing Disorders*, pp. 1–8 (2012).

[10] B. Boren, A. Roginska and B. Gill, Maximum Averaged and Peak Levels of Vocal Sound Pressure, in *Audio Engineering Society Convention 135*, pp. 1–7 (2013).

[11] W. Kim, G. Jang and Y. Young, Microspeaker Diaphragm Optimization for Widening the Operating Frequency Band and Increasing Sound Pressure Level, *IEEE Transactions on Magnetics* **46**, 1, pp. 59–66 (2010).

[12] J. L. Hood, *Valve and Transistors Audio Amplifiers*. Elsevier, Oxford, UK (1997).

[13] P. Sun, J. Park, J. Kwon and S. Hwang, Development of Slim Speaker for Use in Flat TVs, *IEEE Transactions on Magnetics* **48**, 11, pp. 4148–4151 (2012).

[14] H. Takewa, S. Saiki, S. Kano and A. Inaba, Slim-Type Speaker for Flat-Panel Televisions, *IEEE Transactions on Consumer Electronics* **52**, 1, pp. 189–195 (2006).

[15] H. J. Kim, W. S. Yang and K. No, Improvement of Low-Frequency Characteristics of Piezoelectric Speakers Based on Acoustic Diaphragms, *IEEE Transactions on Ultrasonics, Ferroelectrics, and Frequency Control* **59**, 9, pp. 2027–2035 (2012).

[16] X. Branca, B. Allard, X. Lin-Shi and D. Chesneau, Single-Inductor Bipolar-Outputs Converter for the Supply of Audio Amplifiers in Mobile Platforms, *IEEE Transactions on Power Electronics* **28**, 9, pp. 4248–4259 (2013).

[17] C. Kim, G. Moon and S. Han, Voltage Doubler Rectified Boost-Integrated Half Bridge (VDRBHB) Converter for Digital Car Audio Amplifiers, *IEEE Transactions on Power Electronics* **22**, 6, pp. 2321–2330 (2007).

[18] G. Villar-Pique, H. J. Bergveld and E. Alarcon, Survey and Benchmark of Fully Integrated Switching Power Converters: Switched-Capacitor Versus Inductive Approach, *IEEE Transactions on Power Electronics* **28**, 9, pp. 4156–4167 (2013).

[19] Texas Instruments Inc., *TPA2100P1: 19-Vpp Mono Class-D Audio Amplifier for Piezo/Ceramic Speakers* (2008), `http://www.ti.com/lit/ds/symlink/tpa2100p1.pdf`, Dallas, TX, December 2008, Accessed on December 2014.

[20] AKM Semiconductor Inc., *AK7841B: 14-Vpp Mono Class-D Amplifier for Piezo Speaker with Built-in DCDC* (2009), `http://akm.lfchosting.com/datasheets/ak7841_e_13.0.pdf`, Tokyo, Japan, April 2009, Accessed on December 2014.

[21] New Japan Radio Ltd., *NJW1263: Analog Signal Input Class-D Amplifier for Piezo Speaker with DC-DC Converter* (2009), `http://www.njr.com/semicon/PDF/NJW1263_E.pdf`, Tokyo, Japan, Accessed on December 2014.

[22] Maxim Integrated Inc., *MAX9788: 14Vp-p, Class G Ceramic Speaker Driver* (2008), `http://datasheets.maximintegrated.com/en/ds/MAX9788.pdf`, San Jose, CA, May 2008, Accessed on December 2014.

[23] P. Laitinen and J. Maenpaa, Enabling Mobile Haptic Design: Piezoelectric Actuator Technology Properties in Hand Held Devices, in *IEEE International Workshop on Haptic Audio Visual Environments (HAVE) and their Applications*, pp. 40–43 (2006).

[24] B. Hofer, *Guidelines for Measuring Audio Power Amplifier Performance* (2001), `http://www.ti.com`, texas Instruments Inc.

[25] B. Hofer, *Measuring Switch-mode Power Amplifiers* (2003), `http://www.ap.com`, audio Precision Inc.

[26] B. Metzler, *Audio Measurement Handbook* (1993), `http://www.ap.com`, audio Precision Inc.

[27] B. Bauer and E. Torick, Researches in Loudness Measurement, *IEEE Transactions on Audio and Electroacoustics* **14**, 3, pp. 141–151 (1966).

[28] S. El-Hamamsy, Design of High-efficiency RF Class-D Power Amplifier, *IEEE Transactions on Power Electronics* **9**, 3, pp. 297–308 (1994).

[29] S. Burrow and D. Grant, Efficiency of Low Power Audio Amplifiers and Loudspeakers, *IEEE Transactions on Consumer Electronics* **47**, 3, pp. 622–630 (2001).

[30] B. Ducharne, L. Garbuio, M. Lallart, D. Guyomar, G. Sebald and J. Gauthier, Nonlinear Technique for Energy Exchange Optimization in Piezoelectric Actuators, *IEEE Transactions on Power Electronics* **28**, 8, pp. 3941–3948 (2013).

[31] T. Andersen, L. Huang, M. A. E. Andersen and O. C. Thomsen, Efficiency of Capacitively Loaded Converters, in *IEEE Industrial Electronics Society Conference (IECON)*, pp. 368–373 (2012).

[32] D. Nielsen, A. Knott and M. A. E. Andersen, A High-voltage Class D Audio Amplifier for Dielectric Elastomer Transducers, in *IEEE Applied Power Electronics Conference and Exposition (APEC)*, pp. 3278–3283 (2014).

[33] H. Ma, R. Van de Zee and B. Nauta, An Integrated 80V 45W Class-D Power Amplifier with Optimal-Efficiency-Tracking Switching Frequency Regulation, in *IEEE International Solid-State Circuits Conference (ISSCC) Dig. of Tech. Papers*, pp. 286–288 (2014).

[34] D. Nielsen, A. Knott and M. A. E. Andersen, Driving Electrostatic Transducers, in *Audio Engineering Society Convention 134*, pp. 1–7 (2013).

[35] C. Mohan and P. M. Furth, A 16-Ω Audio Amplifier With 93.8-mW Peak Load Power and 1.43-mW Quiescent Power Consumption, *IEEE Transactions on Circuits and Systems II* **59**, 3, pp. 133–137 (2012).

[36] V. Dhanasekaran, J. Silva-Martinez and E. Sanchez-Sinencio, Design of Three-Stage Class-AB 16-Ω Headphone Driver Capable of Handling Wide Range of Load Capacitance, *IEEE Journal of Solid-State Circuits* **44**, 6, pp. 1734–1744 (2009).

[37] F. H. Raab, Average Efficiency of Class-G Power Amplifiers, *IEEE Transactions on Consumer Electronics* **32**, 2, pp. 145–150 (1986).

[38] A. Lollio, G. Bollati and R. Castello, A Class-G Headphone Amplifier in 65 nm CMOS Technology, *IEEE Journal of Solid-State Circuits,* **45**, 12, pp. 2530–2542 (2010).

[39] S. Galal, H. Zheng, K. Abdelfattah, V. Chandrasekhar, I. Mehr, A. J. Chen, J. Platenak, N. Matalon and T. L. Brooks, A 60 mW Class-G Stereo Headphone Driver for Portable Battery-Powered Devices, *IEEE Journal of Solid-State Circuits,* **47**, 8, pp. 1921–1934 (2012).

[40] B. Bhamidipati, A. I. Colli-Menchi and E. Sanchez-Sinencio, Low Power Complementary Metal-oxide Semiconductor Class-G Audio Amplifier with Gradual Power Supply Switching, *IET Circuits, Devices, and Systems* **9**, 4, pp. 256–264 (2015).

[41] H. Nam, Y. Ahn and J. Roh, 5-V Buck Converter Using 3.3-V Standard CMOS Process With Adaptive Power Transistor Driver Increasing Efficiency and Maximum Load Capacity, *IEEE Transactions on Power Electronics* **27**, 1, pp. 463–471 (2012).

[42] W. Liou, M. Yehm and Y. L. Kuo, A High Efficiency Dual-Mode Buck Converter IC for Portable Applications, *IEEE Transactions on Power Electronics* **23**, 2, pp. 667–677 (2008).

[43] P. Y. Wu, S. Y. S. Tsui and P. K. T. Mok, Area- and Power-Efficient Monolithic Buck Converters With Pseudo-Type III Compensation, *IEEE J. of Solid-State Circuits* **45**, 8, pp. 1446–1455 (2010).

[44] M. Teplechuk, T. Gribben and C. Amadi, True Filterless Class-D Audio Amplifier, *IEEE Journal of Solid-State Circuits* **46**, 12, pp. 2784–2793 (2011).

[45] V. M. E. Antunes, V. F. Pires and J. F. A. Silva, Narrow Pulse Elimination PWM for Multilevel Digital Audio Power Amplifiers Using Two Cascaded H-Bridges as a Nine-Level Converter, *IEEE Transactions on Power Electronics* **22**, 2, pp. 425–434 (2007).

[46] Texas Instruments Inc., *AN-1737 Managing EMI in Class D Audio Applications* (2013), `http://www.ti.com/lit/an/snaa050a/snaa050a.pdf`, Dallas, TX, May 2013, Accessed on December 2014.

[47] X. Ming, Z. Chen, Z. Zhou and B. Zhang, An Advanced Spread Spectrum Architecture Using Pseudorandom Modulation to Improve EMI in Class D Amplifier, *IEEE Transactions on Power Electronics* **26**, 2, pp. 638–646 (2011).

[48] M. Yeh, W. Liou, H. Hsieh and Y. Lin, An Electromagnetic Interference (EMI) Reduced High-Efficiency Switching Power Amplifier, *IEEE Transactions on Power Electronics* **25**, 3, pp. 710–718 (2010).

[49] J. Liu, P. Wang and T. Kuo, A Current-Mode DC-DC Buck Converter with Efficiency-Optimized Frequency Control and Reconfigurable Compensation, *IEEE Transactions on Power Electronics* **27**, 2, pp. 869–880 (2012).

[50] Y. Ren, M. Xu, J. Zhou and F. Lee, Analytical Loss Model of Power MOSFET, *IEEE Transactions on Power Electronics* **21**, 2, pp. 310–319 (2006).

[51] J. Wang, R. Li and H. Chung, An Investigation Into the Effects of the Gate Drive Resistance on the Losses of the MOSFET-Snubber-Diode Configuration, *IEEE Transactions on Power Electronics* **27**, 5, pp. 2657–2672 (2012).

[52] J. Sun, M. Xu, Y. Ren and F. Lee, Light-Load Efficiency Improvement for Buck Voltage Regulators, *IEEE Transactions on Power Electronics* **24**, 3, pp. 742–751 (2009).

[53] M. Berkhout, A Class D Output Stage with Zero Dead Time, in *IEEE International Solid-State Circuits Conference (ISSCC) Digital of Technical Papers*, pp. 134–135 (2003).

[54] D. G. Holmes and T. A. Lipo, *Pulse Width Modulation for Power Converters*. John Wiley & Sons Inc., New York, USA, (2003).

[55] M. T. Tan, J. S. Chang, H. C. Chua and B. H. Gwee, An Investigation into the Parameters Affecting Total Harmonic Distortion in Low-Voltage Low-Power Class-D Amplifiers, *IEEE Transactions on Circuits and Systems I* **50**, 10, pp. 1304–1315 (2003).

[56] B. Forejt, V. Rentala, J. D. Arteaga and G. Burra, A 700+-mW Class D Design with Direct Battery Hookup in a 90-nm Process, *IEEE Journal of Solid-State Circuits* **40**, 9, pp. 1880–1887 (2005).

[57] Texas Instruments Inc., *LM4666 Boomer: Filterless High Efficiency Stereo 1.2W Switching Audio Amplifier* (2013), `http://www.ti.com/lit/ds/symlink/lm4666.pdf`, Dallas, TX, May 2013, Accessed on December 2014.

[58] P. P. Siniscalchi and R. K. Hester, A 20W/channel Class-D Amplifier with Significantly Reduced Common-mode Radiated Emissions, in *IEEE International Solid-State Circuits Conference (ISSCC) Digitial of Technical Papers*, pp. 448–449 (2009).

[59] W. Shu and J. Chang, Power Supply Noise in Analog Audio Class D Amplifiers, *IEEE Transactions on Circuits and Systems I* **56**, 1, pp. 84–96 (2009).

[60] T. Ge, J. S. Chang and W. Shu, PSRR of Bridge-Tied Load PWM Class D Amps, in *IEEE International Symposium on Circuits and Systems (ISCAS)*, pp. 284–287 (2008).

[61] P. Allen and D. Holberg, *CMOS Analog Circuit Design.* Oxford University Press, New York (2002).

[62] Cirrus Logic Inc., *CS35L00: 2.8 W Mono Class-D Audio Amplifier with Low Idle Current* (2011), `http://www.cirrus.com/en/pubs/proDatasheet/CS35L00_PP1.pdf`, Austin, TX, February 2011, Accessed on December 2014.

[63] Analog Devices Inc., *SSM2306: 2W Filterless Class-D Stereo Audio Amplifier* (2007), `http://www.analog.com/static/imported-files/data_sheets/SSM2306.pdf`, Norwood, MA, May 2007, Accessed on December 2014.

[64] NXP Semiconductors N.V., *SA58672: 3 W Mono Class-D Audio Amplifier* (2009), `http://www.nxp.com/documents/data_sheet/SA58672.pdf`, Eindhoven, Netherlands, June 2009, Accessed on December 2014.

[65] Fairchild Semiconductor Corp., *FAB3103: 2.3 Watt Class-D Audio Amplifier with Integrated Boost Regulator and Automatic Gain Control* (2012), `https://www.fairchildsemi.com/datasheets/FA/FAB3103.pdf`, San Jose, CA, July 2012, Accessed on December 2014.

[66] R. Erickson and D. Maksimovic, *Fundamentals of Power Electronics.* Kluwer Academic Publishers, New York (2000).

[67] P. P. Vaidyanathan, Generalizations of the Sampling Theorem: Seven Decemberades after Nyquist, *IEEE Transactions on Circuits and Systems I* **48**, 9, pp. 1094–1109 (2001).

[68] J. Yu, M. T. Tan, S. M. Cox and W. Goh, Time-Domain Analysis of Intermodulation Distortion of Closed-Loop Class-D Amplifiers, *IEEE Transactions on Power Electronics* **27**, 5, pp. 2453–2461 (2012).

[69] L. Guo, T. Ge and J. Chang, Intermodulation Distortions of Bang-Bang Control Class D Amplifiers, *IEEE Transactions on Power Electronics* **29**, 12, pp. 6604–6614 (2014).

[70] M. Hoyerby and M. A. E. Andersen, Carrier Distortion in Hysteretic Self-Oscillating Class-D Audio Power Amplifiers: Analysis and Optimization, *IEEE Transactions on Power Electronics* **24**, 3, pp. 714–729 (2009).

[71] A. Colli-Menchi, J. Torres and E. Sanchez-Sinencio, A Feed-forward Power-supply Noise Cancellation Technique for Single-Ended Class-D Audio Amplifiers, *IEEE Journal of Solid-State Circuits* **49**, 3, pp. 718–728 (2014).

[72] Y. Choi, W. Tak, Y. Yoon, J. Roh, S. Kwon and J. Koh, A 0.018% THD+N, 88-dB PSRR PWM Class-D Amplifier for Direct Battery Hookup, *IEEE Journal of Solid-State Circuits* **47**, 2, pp. 454–463 (2012).

[73] C. K. Lam, M. T. Tan, S. M. Cox and K. S. Yeo, Class-D Amplifier Power Stage With PWM Feedback Loop, *IEEE Transactions on Power Electronics* **28**, 8, pp. 3870–3881 (2013).

[74] J. Chang, M. Tan, Z. Cheng and Y. Tong, Analysis and Design of Power Efficient Class D Amplifier Output Stages, *IEEE Transactions on Circuits and Systems I* **47**, 6, pp. 897–902 (2000).

[75] M. Wang, X. Jiang, J. Song and T. L. Brooks, A 120 dB Dynamic Range 400 mW Class-D Speaker Driver with Fourth-Order PWM Modulator, *IEEE Journal of Solid-State Circuits* **45**, 8, pp. 1427–1435 (2010).

[76] M. Kinyua, R. Wang and E. Soenen, Integrated 105 dB SNR, 0.0031% THD+N Class-D Audio Amplifier with Global Feedback and Digital Control in 55 nm CMOS, *IEEE Journal of Solid-State Circuits* **50**, 8, pp. 1764–1771 (2015).

[77] E. Gaalaas, B. Y. Liu, N. Nishimura, R. Adams and K. Sweetland, Integrated Stereo Delta-Sigma Class-D Audio Amplifier, *IEEE Journal of Solid-State Circuits* **40**, 12, pp. 2388–2397 (2005).

[78] M. A. Rojas-Gonzalez and E. Sanchez-Sinencio, Low-Power High-Efficiency Class D Audio Power Amplifiers, *IEEE Journal of Solid-State Circuits* **44**, 12, pp. 3272–3284 (2009).

[79] J. Torres, A. Colli-Menchi, M. A. Rojas-Gonzalez and E. Sanchez-Sinencio, A Low-Power High-PSRR Clock-Free Current-Controlled Class-D Audio Amplifier, *IEEE Journal of Solid-State Circuits* **46**, 7, pp. 1553–1561 (2011).

[80] M. A. Rojas-González and E. Sánchez-Sinencio, Design of a Class D Audio Amplifier IC Using Sliding Mode Control and Negative Feedback, *IEEE Transactions on Consumer Electronics* **53**, 2, pp. 609–617 (2007).

[81] J. Torres, A. Colli-Menchi, M. A. Rojas-Gonzalez and E. Sanchez-Sinencio, A 470 μW Clock-Free Current-Controlled Class D Amplifier with 0.02% THD+N and 82 dB PSRR, in *IEEE European Solid-State Circuits Conference (ESSCIRC)*, pp. 326–329 (2010).

[82] M. Rojas-Gonzalez and E. Sanchez-Sinencio, Two Class-D Audio Amplifiers with 89/901 mW of Quiescent Power, in *IEEE International Solid-State Circuits Conference (ISSCC) Dig. of Tech. Papers*, pp. 450–451,451a (2009).

[83] J. Lu and R. Gharpurey, Design and Analysis of a Self-Oscillating Class D Audio Amplifier Employing a Hysteretic Comparator, *IEEE Journal of Solid-State Circuits* **46**, 10, pp. 2336–2349 (2011).

[84] R. Schreier and G. C. Temes, *Understanding Delta-Sigma Data Converters.* Wiley-IEEE, New Jersey (2004).

[85] M. K. Alghamdi and A. A. Hamoui, A Spurious-Free Switching Buck Converter Achieving Enhanced Light-Load Efficiency by Using a $\Delta\Sigma$ — Modulator Controller With a Scalable Sampling Frequency, *IEEE Journal of Solid-State Circuits* **47**, 4, pp. 841–851 (2012).

[86] B. Bhattacharyya, Describing-Function Expressions for Sine-type Functional Non-linearity in Feedback Control Systems, *Proceedings of the IEE* **108-B**, 41, pp. 529–534 (1961).

[87] B. Labbe, B. Allard, X. Lin-Shi and D. Chesneau, An Integrated Sliding-Mode Buck Converter with Switching Frequency Control for Battery-Powered Applications, *IEEE Transactions on Power Electronics* **28**, 9, pp. 4318–4326 (2013).

[88] S. Tan, Y. M. Lai, C. K. Tse and M. Cheung, Adaptive Feedforward and Feedback Control Schemes For Sliding Mode Controlled Power Converters, *IEEE Transactions on Power Electronics* **21**, 1, pp. 182–192 (2006).

[89] M. Hoyerby and M. A. E. Andersen, Ultrafast Tracking Power Supply With Fourth-Order Output Filter and Fixed-Frequency Hysteretic Control, *IEEE Transactions on Power Electronics* **23**, 5, pp. 2387–2398 (2008).

[90] S. S. Ng, K. Lin, K. Chen and Y. Chen, A 94 % Efficiency Near-Constant Frequency Self-Oscillating Class-D Audio Amplifier with Voltage Control Resistor, in *IEEE Internatinoal Symposium on Circuits and Systems (ISCAS)*, pp. 602–605 (2013).

[91] D. Nielsen, A. Knott and M. A. E. Andersen, Hysteretic Self-oscillating Bandpass Current Mode Control for Class D Audio Amplifiers Driving Capacitive Transducers, in *IEEE Energy Conversion Congress and Exposition (ECCE) Asia*, pp. 971–975 (2013).

[92] V. Utkin, J. Guldner and J. Shi, *Sliding Mode Control in Electro-Mechanical Systems.* CRC Press, London (2009).

[93] S. K. Mazumder and S. L. Kamisetty, Design and Experimental Validation of a Multiphase VRM Controller, *IEE Proceedings on Electric Power Applications* **152**, 5, pp. 1076–1084 (2005).

[94] F. B. Cunha, D. J. Pagano and U. F. Moreno, Sliding Bifurcations of Equilibria in Planar Variable Structure Systems, *IEEE Transactions on Circuits and Systems I* **50**, 8, pp. 1129–1134 (2003).

[95] F. Su, W. H. Ki and C. Y. Tsui, Ultra Fast Fixed-Frequency Hysteretic Buck Converter With Maximum Charging Current Control and Adaptive Delay Compensation for DVS Applications, *IEEE Journal of Solid-State Circuits* **43**, 4, pp. 815–822 (2008).

[96] P. Hazucha, G. Schrom, J. Hahn, B. A. Bloechel, P. Hack, G. E. Dermer, S. Narendra, D. Gardner, T. Karnik, V. De and S. Borkar, A 233-MHz 80%-87% Efficient Four-phase DC-DC Converter Utilizing Air-core Inductors on Package, *IEEE Journal of Solid-State Circuits* **40**, 4, pp. 838–845 (2005).

[97] R. G. Eschauzier, L. P. Kerklaan and J. H. Huijsing, A 100-MHz 100-dB Operational Amplifier with Multipath Nested Miller Compensation Structure, *IEEE Journal of Solid-State Circuits* **27**, 12, pp. 1709–1717 (1992).

[98] X. Fan, C. Mishra and E. Sanchez-Sinencio, Single Miller Capacitor Frequency Compensation Technique for Low-power Multistage Amplifiers, *IEEE Journal of Solid-State Circuits* **40**, 3, pp. 584–592 (2005).

[99] R. G. Eschauzier and J. H. Huijsing, *Frequency Compensation Techniques for Low-Power Operational Amplifiers.* Dordrecht, The Netherlands, Kluwer Academic Publishers (1995).

[100] P. Gray, P. Hurst, S. Lewis and R. Meyer, *Analysis and Design of Analog Integrated Circuits.* John Wiley & Sons, New Jersey (2001).

[101] S. Franco, *Design with Operational Amplifiers and Analog Integrated Circuits.* McGraw-Hill Higher Education, New York (2002).

[102] G. M. Yin, F. Eynde and W. Sansen, A High-speed CMOS Comparator with 8-b Resolution, *IEEE Journal of Solid-State Circuits* **27**, 2, pp. 208–211 (1992).

[103] F. Koeslag, H. D. Mouton and J. Beukes, Analytical Modeling of the Effect of Nonlinear Switching Transition Curves on Harmonic Distortion in Class D Audio Amplifiers, *IEEE Transactions on Power Electronics* **28**, 1, pp. 380–389 (2013).

[104] J. S. Choi and K. Lee, Design of CMOS Tapered Buffer for Minimum Power-Delay Product, *IEEE Journal of Solid-State Circuits* **29**, 9, pp. 1142–1145 (1994).

[105] K. H. Chen and Y. S. Hsu, A High-PSRR Reconfigurable Class-AB/D Audio Amplifier Driving a Hands-Free/Receiver 2-in-1 Loudspeaker, *IEEE Journal of Solid-State Circuits* **47**, 11, pp. 2586–2603 (2012).

[106] Y. Lin, C. Lee and Y. Tzou, Architecture Implementation of Class-D Amplifiers Using Digital-Controlled Multiphase-Interleaved PWM Technique, in *CES/IEEE International Power Electronics and Motion Control Conference (IPEMC)*, Vol. 2, pp. 1–6 (2006).

[107] P. Xu, J. Wei and F. C. Lee, Multiphase Coupled-Buck Converter-A Novel High Efficient 12 V Voltage Regulator Module, *IEEE Transactions on Power Electronics* **18**, 1, pp. 74–82 (2003).

[108] I. Lee, S. Cho and G. Moon, Interleaved Buck Converter Having Low Switching Losses and Improved Step-Down Conversion Ratio, *IEEE Transactions on Industrial Electronics* **27**, 8, pp. 3664–3675 (2012).

[109] H. N. Nagaraja, D. K. Kastha and A. Petra, Design Principles of a Symmetrically Coupled Inductor Structure for Multiphase Synchronous Buck Converters, *IEEE Transactions on Industrial Electronics* **58**, 3, pp. 988–997 (2011).

[110] A. Nagari, E. Allier, F. Amiard, V. Binet and C. Fraisse, An 8Ω 2.5W 1% THD 104dB(A)-Dynamic-Range Class-D Audio Amplifier With Ultra-Low EMI System and Current Sensing for Speaker Protection, *IEEE Journal of Solid-State Circuits* **47**, 12, pp. 3068–3080 (2012).

[111] A. M. Patel and M. Ferdowsi, Advanced Current Sensing Techniques for Power Electronic Converters, in *IEEE Vehicle Power and Propulsion Conference (VPPC)*, pp. 524–530 (2007).

[112] H. P. Forghani-zadeh and G. A. Rincon-Mora, Current-Sensing Techniques for DC-DC Converters, in *IEEE Int. Midwest Symposium on Circuits and Systems (MWSCAS)*, pp. 577–580 (2002).

[113] L. Guo, T. Ge and J. Chang, An Ultra-low-power Overcurrent Protection Circuit for Micropower Class-D Amplifiers, *IEEE Transactions on Circuits and Systems II* **PP**, 99, pp. 1–5 (2015).

[114] M. Ryu, J. Kim, J. Baek and H. Kim, New Multi-channel LEDs Driving Methods using Current Transformer in Electrolytic Capacitor-less AC-

DC Drivers, in *IEEE Applied Power Electronics Conference and Exposition (APEC)*, pp. 2361–2367 (2012).

[115] A. T. Mozipo, Analysis of a Current Sensing and Reporting Tolerance in an Inductor DCR Sense Topology, in *IEEE Applied Power Electronics Conference and Exposition (APEC)*, pp. 1502–1506 (2011).

[116] B. Putzeys, *Digital Audio's Final Frontier.* IEEE Spectrum, pp. 34–41 (2003).

[117] D. Grant, Y. Darroman and S. Burrow, Class-D Amplification Combined with Switched-Mode Power Conditioning — The Route to Long Battery Life, *IEEE Transactions on Consumer Electronics* **48**, 3, pp. 677–683 (2002).

[118] X. Jiang, J. Song, M. Wang, J. Chen, H. Zheng, S. Galal, K. Abdelfattah and T. L. Brooks, A 100 dB DR Ground-Referenced Single-Ended Class-D Amplifier in 65 nm CMOS, in *IEEE Symposium on VLSI Circuits (VLSIC)*, pp. 58–59 (2011).

[119] A. Huffenus, G. Pillonnet, N. Abouchi, F. Goutti, V. Rabary and R. Cittadini, A High PSRR Class-D Audio Amplifier IC Based on a Self-Adjusting Voltage Reference, in *IEEE European Solid-State Circuits Conference (ESSCIRC)*, pp. 118–121 (2010).

[120] A. Hastings, *The Art of Analog Layout.* Prentice-Hall, New Jersey (2001).

[121] S. Sanders, On Limit Cycles and the Describing Function Method in Periodically Switched Circuits, *IEEE Transactions on Circuits and Systems I* **40**, 9, pp. 564–572 (1993).

[122] L. R. Carley, A Noise-Shaping Coder Topology for 15+ Bit Converters, *IEEE Journal of Solid-State Circuits* **24**, 2, pp. 267–273 (1989).

[123] P. Muggler, W. Chen, C. Jones, P. Dagli and N. Yazdi, A Filter Free Class D Audio Amplifier with 86% Power Efficiency, in *IEEE International Symposium on Circuits and Systems*, pp. 1036–1039 (2004).

[124] J. Baker, *CMOS: Circuit Design, Layout, and Simulation.* Wiley-IEEE, New Jersey (2008).

[125] B. Sahu and G. A. Rincon-Mora, A Low Voltage, Dynamic, Noninverting, Synchronous Buck-Boost Converter for Portable Applications, *IEEE Transactions on Power Electronics* **19**, 2, pp. 443–452 (2004).

[126] L. Huang, Z. Zhang and M. A. E. Andersen, A Review of High Voltage Drive Amplifiers for Capacitive Actuators, in *IEEE International Universities Power Engineering Conference (UPEC)*, pp. 1–6 (2012).

[127] S. I. Furuya, T. Maruhashi, Y. Izuno and M. Nakaoka, Load-Adaptive Frequency Tracking Control Implementation of Two-Phase Resonant Inverter for Ultrasonic Motor, *IEEE Transactions on Power Electronics* **7**, 3, pp. 542–550 (1992).

[128] A. Colli-Menchi and E. Sanchez-Sinencio, A High-Efficiency Self-Oscillating Class-D Amplifier for Piezoelectric Speakers, *IEEE Transactions on Power Electronics,* **30**, 9, pp. 5125–5135 (2015).

[129] B. Serneels, M. Steyaert and W. Dehaene, A High Speed, Low Voltage to High Voltage Level Shifter in Standard 1.2V 0.13 μm CMOS, in *IEEE Int. Conference on Electronics, Circuits and Systems (ICECS)*, pp. 668–671 (2006).

[130] H. K. Khalil, *Nonlinear Systems.* Prentice-Hall, New Jersey (2001).

Index

Printed in the United States
By Bookmasters